Progress in Probability

Volume 33

Progress in Probability
Volume 33

Seminar on Stochastic Processes, 1992

E. Çinlar
K. L. Chung
M. J. Sharpe
Editors

R. F. Bass
K. Burdzy
Managing Editors

Springer Science+Business Media, LLC

E. Çınlar
Dept. of Civil Engineering
and Operations Research
Princeton University
Princeton, NJ 08544

K. L. Chung
Dept. of Mathematics
Stanford University
Stanford, CA 94305

R.F. Bass
K. Burdzy
(Managing Editors)
Dept. of Mathematics
University of Washington
Seattle, WA 98195

M. J. Sharpe
Dept. of Mathematics
University of California-San Diego
La Jolla, CA 92093

Library of Congress Cataloging-in-Publication Data
Seminar on Stochastic Processes (12th : University of Washington)
Seminar on Stochastic Processes, 1992 / E. Çınlar, K. L. Chung, M. J. Sharpe, editors ; R. F. Bass, K. Burdzy, managing editors.
p. cm. -- (Progress in probability ; v. 33)
Seminar held at the University of Washington Mar. 26-28, 1992.
Includes bibliographical references.
ISBN 978-1-4612-6714-0 ISBN 978-1-4612-0339-1 (eBook)
DOI 10.1007/978-1-4612-0339-1
acid-free)
1. Stochastic processes--Congresses. I. Çınlar, E. (Erhan), 1941- . II. Chung, Kai Lai, 1917- . III. Sharpe, M. J., 1941- . IV. Title. V. Series: Progress in probability : 33.
QA274.A1S44 1992 93-22016
519.2--dc20 CIP

Printed on acid-free paper.

ISBN 978-1-4612-6714-0
Camera-ready copy prepared by the Authors in TEX.

9 8 7 6 5 4 3 2 1

Contents

FOREWORD

The 1992 Seminar on Stochastic Processes was held at the University of Washington from March 26 to March 28, 1992. This was the twelfth in a series of annual meetings which provide researchers with the opportunity to discuss current work on stochastic processes in an informal and enjoyable atmosphere. Previous seminars were held at Northwestern University, Princeton University, University of Florida, University of Virginia, University of California, San Diego, University of British Columbia and University of California, Los Angeles. Following the successful format of previous years, there were five invited lectures, delivered by R. Adler, R. Bañuelos, J. Pitman, S. J. Taylor and R. Williams, with the remainder of the time being devoted to informal communications and workshops on current work and problems. The enthusiasm and interest of the participants created a lively and stimulating atmosphere for the seminar. A sample of the research discussed there is contained in this volume.

The 1992 Seminar was made possible through the support of the National Science Foundation, the National Security Agency, the Institute of Mathematical Statistics and the University of Washington. We extend our thanks to them and to the publisher Birkhäuser Boston for their support and encouragement.

Richard F. Bass
Krzysztof Burdzy

Seattle, 1992

SUPERPROCESS LOCAL AND INTERSECTION LOCAL TIMES AND THEIR CORRESPONDING PARTICLE PICTURES

Robert J. Adler

Faculty of Industrial Engineering & Management
Technion – Israel Institute of Technology

Abstract

The super Brownian motion has a local time up to three dimensions, an intersection (over disjoint time sets) local time up to dimension seven, and a renormalised self-intersection local time up to dimension five.

The branching Brownian motions that, in the infinite density limit, provide a particle picture for the super Brownian motion have a local time in only one dimension, an intersection local time up to three dimensions, and a renormalised self-intersection local time only in dimensions one and two.

This paper is primarily concerned with the "dimension gap" between the particle picture and the superprocess, and shows how to explain this gap in terms of weak convergence of functionals of the finite system to functionals on the superprocess. The first section gives a detailed review of the recent literature on local and intersection local times for super Brownian and super stable processes.

AMS 1991 subject classifications: Primary, 60J55, 60H15; Secondary, 60F17, 60F25.

Key words and phrases: Superprocesses, local time, intersection local time.

1. Research supported in part by US-Israel Binational Science Foundation (89-298), Israel Academy of Sciences (702-90), and Office of Naval Research (N00014/89/J/1870).

0. INTRODUCTION

This paper is concerned with the local time and intersection local time of superprocesses. In particular, we are interested in how these relate to certain functionals of the branching particle system whose high density limit is the superprocess.

The first section is a mini-review of the recent literature on the local times of superprocesses. It contains no new results (although one or two "well known" facts may appear here in print for the first time), and so can be skipped by a reader already familiar with this material.

Section 2 contains the new results, as well as summarising briefly some previous, qualitatively different, results that arise when the questions raised in this paper are applied to the Brownian density process.

Section 3 contains some moment formulae for the finite systems of branching particles mentioned above. These are closely related to Dynkin's (1988c) diagrams for calculating the mixed moments of the superprocesses, and should be of independent interest. In the same vein that we believe the particle picture adds probabilistic insight into the structure of superprocess local time, we believe moment formulae for the particle picture add further understanding to Dynkin's analytically derived formulae.

Sections 4 and 5 are devoted to proofs.

1. LOCAL TIMES

(a) Superprocesses and particle pictures. We start with a parameter $\mu > 0$ that will eventually become large, and a finite measure m on $\Re^d$, $d \geq 1$, of total mass $|m| = m(\Re^d)$. With these we generate a particle picture based on $K = [\mu \cdot |m|]$ initial particles which, at time zero, are independently distributed in $\Re^d$, according to the probability measure $m(\cdot)/|m|$. Later on we shall occasionally require that m is absolutely continuous with respect to Lebesgue measure. We define a branching rate $\rho > 0$. Each of these K particles follows the path of independent copies of a Markov process B, until time $t = \rho/\mu$.

At time ρ/μ each particle, independently of the others, either dies or splits into two, with probability $\frac{1}{2}$ for each event. The individual particles in the new population then follow independent copies of B, starting at their place of birth, in the interval $[\rho/\mu, 2\rho/\mu)$, and the pattern of alternating critical branching and spatial spreading continues until, with probability one, there are no particles left alive.

The process of interest to us is the measure valued Markov process

$$(1.1) \qquad X_t^\mu(A) = \frac{\{\text{Number of particles in } A \text{ at time } t\}}{\mu},$$

where $A \in \mathcal{B}^d$ = Borel sets in $\Re^d$. Note that, for fixed t and μ, X_t^μ is an atomic measure. Note also that if $\rho = \infty$ there is no branching occuring.

It is now well known that under very mild conditions on B the sequence $\{X^\mu\}_{\mu \geq 1}$ converges weakly, on an appropriate Skorohod space, to a measure valued process which is called the superprocess for B. A precise formulation of this convergence is given in §2. We shall be interested primarily in the cases in which B is either an $\Re^d$ valued Brownian motion with infinitesimal generator Δ, (so that the components of B have variance 2 at time $t = 1$) or an $\Re^d$ valued symmetric stable process whose generator is the fractional Laplacian $\Delta_\alpha := -(-\Delta)^{\alpha/2}$, $\alpha \in (0, 2)$. (cf. Yoshida (1980).) When $\alpha = 2$ we define $\Delta_2 \equiv \Delta$. Set $\rho = 1$. Then in either of these cases the corresponding superprocess, known, respectively, as a super Brownian motion or super stable process, can be defined as the unique solution of the following martingale problem:

For all $\phi \in C_b^2(\Re^d)$, the space of all bounded, continuous, $\Re$ valued functions on $\Re^d$ with continuous first and second order partial derivatives,

$$Z_t(\phi) = X_t(\phi) - m(\phi) - \int_0^t X_s(\Delta_\alpha \phi)\, ds, \tag{1.2a}$$

is a continuous square integrable martingale such that $Z_0 = 0$ and

$$\langle Z_.(\phi) \rangle_t = \int_0^t X_s(\phi^2)\, ds. \tag{1.2b}$$

(We have taken the obvious liberty here of denoting integration via $\int \phi(x)\, \nu(dx) = \nu(\phi)$ for a measure ν. Later, without further comment, we shall also write this as $\langle \phi, \nu \rangle$.)

It can be shown that the solution of this martingale problem is also the unique solution of the stochastic partial differential equation (SPDE)

$$\dot{X}_t = \Delta_\alpha^* X_t + \dot{Z}_t. \tag{1.3}$$

We shall call the system of branching particles described in the first two paragraphs the "particle picture" for the measure valued solution of (1.2). A more precise description of this system is given in the following section.

When $\rho = \infty$ the situation is a little different, and X_t^μ, as defined above, converges with probability one to the absolutely continuous measure whose density $x_t(u)$ satisfies the deterministic PDE $\partial x_t(u)/\partial t = \Delta_\alpha x_t(u)$, with initial condition $x_0 = m$. (This is just the strong law of large numbers.) In this case we look at the processes

$$\eta_t^\mu(\phi) = \mu^{1/2} \{X_t^\mu(\phi) - EX_t^\mu(\phi)\}, \tag{1.4}$$

where $\phi \in \mathcal{S}_d = \mathcal{S}(\Re^d)$, the Schwartz space of infinitely differentiable functions on $\Re^d$ decreasing rapidly at infinity, so that we think of η as a random (Schwartz) distribution rather than as a (signed) measure. The sequence $\{\eta^\mu\}_{\mu\geq 1}$ also converges weakly. Its limit is called the density process corresponding to B, and it satisfies the SPDE

$$(1.5) \qquad \left(\frac{\partial \eta}{\partial t}\right)(\phi) \;=\; \eta(\Delta_\alpha \phi) \;+\; \sqrt{2} W_t(\Delta_{\alpha/2}\phi),$$

where W is a mean zero, space-time Gaussian process with covariance functional

$$(1.6) \quad E\,W(\phi_1 \times \psi_1) W(\phi_2 \times \psi_2) \;=\; \int_0^\infty \phi_1(t)\phi_2(t)\,dt \int_{\Re^d} \psi_1(x)\,\psi_2(x)\,dx.$$

with $\phi_i \in \mathcal{S}_1$ and $\psi_i \in \mathcal{S}_d$. In this case the limit process η is Gaussian.

Although our main interest will be with superprocesses rather than density processes, we shall have occasion to use the latter to help sort out certain "intuitive" arguments.

Details of the above described weak convergences can be found in Walsh (1986), where one can also find a precise formulation of the SPDE's. Walsh gives a full proof of convergence in the Brownian density case, while the critical branching case can be found, for example, in Watanabe's (1968) original paper or in Ethier and Kurtz (1986).

(b) Superprocess local and intersection local times. The occupation time process for a superprocess X is a further measure valued process defined by

$$(1.7) \qquad L_t(B) = \int_0^t X_s(B)\,ds.$$

This is clearly well defined for every Borel $B \in \Re^d$. The local time process is, formally, obtained by putting $B = \{x\}$ in (1.7). More precisely, let f be a positive C^∞ function supported in the unit d-ball, such that $\int_{\Re^d} f(x)dx = 1$. For each $\epsilon > 0$ set

$$(1.8) \qquad f_\epsilon(x) \;= \epsilon^{-d} f(x/\epsilon),$$

and define

$$(1.9) \qquad L_t^x \;=\; \lim_{\epsilon\to 0} \int_0^t X_s(f_\epsilon(\cdot - x))\,ds.$$

The limit here, when it exists, may be taken in distribution (Iscoe (1986a,b), Fleischmann (1988)) or in $\mathcal{L}^2$ (Dynkin (1988c)), and is independent of the choice of f. An extremely informative paper on the construction of local time is Sugitani (1987).

In order to be more precise about when the limit exists, we shall restrict ourselves to super Brownian and super stable processes. In these cases, not only can one establish existence results, but there are nice evolution equations describing the temporal evolution of the local time. To state these, we need to introduce, for each $\lambda \geq 0$, the Green's functions

$$G_\alpha^\lambda(x,y) = \int_0^\infty e^{-\lambda t} p_t^\alpha(x,y)\,dt, \tag{1.10}$$

corresponding to the transition probabilities

$$p_t^\alpha(x,y) = p_t^\alpha(x-y) = \frac{1}{(2\pi)^d}\int_{\Re^d} \exp\big(-ip\cdot(x-y) - t\|p\|^\alpha\big)\,dp,$$

which, when $\alpha = 2$, is equivalent to Brownian transition probability

$$p_t^{(2)}(x,y) = p_t(x,y) = p_t(x-y) = \frac{1}{(4\pi t)^{d/2}} e^{-\|x-y\|^2/4t}.$$

Green's functions always exist when $\lambda > 0$. The $\lambda = 0$ case often requires restrictions on the dimension d. The most important property of the Green's function, as far as we shall be concerned, is that it satisfies the distributional equation

$$(-\Delta_\alpha + \lambda)u = \delta, \tag{1.11}$$

where δ is the Dirac delta function.That is, for every test function $\phi \in \mathcal{S}_d$,

$$\int_{\Re^d} ((-\Delta_\alpha + \lambda)G_\alpha^\lambda)(x)\phi(x)dx = \phi(0).$$

By substituting $\phi = G_\alpha^\lambda$ into the basic SPDE (1.3) and passing to a formal limit in (1.9) one gets the evolution equation of the following theorem. The (not-insubstantial) justification of this substitution is due to Adler and Lewin (1992). The basic existence result is due to the authors mentioned above.

1.1 THEOREM. *Let X_t be either a super Brownian motion in dimension $d \leq 3$ or a super stable process in dimension $d < 2\alpha$, with $\alpha \in (1,2]$ if $d = 1$. Then the local time L_t^x (1.9) is well defined as an $\mathcal{L}^2$ limit for each $x \in \Re^d$, is independent of the choice of f, and L_t^0 satisfies the following evolution equation for each $\lambda > 0$.*

$$L_t^0 = \langle G_\alpha^\lambda, X_0\rangle - \langle G_\alpha^\lambda, X_t\rangle + \lambda\int_0^t \langle G_\alpha^\lambda, X_s\rangle\,ds + \int_0^t \langle G_\alpha^\lambda, Z(ds)\rangle. \tag{1.12}$$

An identical equation holds for L_t^x if $G_\alpha^\lambda(\cdot)$ is replaced by $G_\alpha^\lambda(\cdot - x)$.

(Dynkin (1988c) has a different type of representation for superprocess local time, and shows that

$$\begin{aligned} L_t^x &= \int_0^t \Big\langle \int_0^s p_{s-u}^\alpha(\cdot, x)du,\, Z(ds)\Big\rangle \\ &= \int_{s=0}^t \int_{u=0}^s \int_{\Re^d} p_{s-u}^\alpha(y, x)\, Z(ds, dy)du. \end{aligned} \tag{1.13}$$

Dynkin's representation comes primarily from Fourier type considerations, although it is also obtainable by explicitly solving the SPDE (1.3). Despite its relatively neat and succint form, it will turn out that (1.12) is more suited for a study based on stochastic analytic techniques and for the particle picture approximations that we shall look at below. This will be particularly so when we turn, soon, to the study of renormalised intersection local times.)

Krone (1991) describes the properties of L_t^x as a function of both t and x. What is important for us at this stage is the dimensions for which the local time exists, and the structure of the evolution equation (1.12).

We now turn to the issue of (self)-intersection local time (SILT). At a heuristic level, this is a set indexed functional of the form

$$SILT(B) = \int_B dsdt \int_{\Re^d \times \Re^d} \delta(x-y)\, X_s(dx)X_t(dy), \tag{1.14}$$

where B is a finite rectangle in $[0,\infty) \times [0,\infty)$ and δ is the Dirac delta function. A more precise definition involves a limit as in the case of the regular local time. There are two qualitatively different cases that have to be considered in a rigorous formulation. The first, and by far the simpler case, arises when the set B does not intersect the diagonal $D = \{(t,t) : t \geq 0\}$. This case has been considered in detail in Dynkin (1988c) for quite general superprocesses. Theorem 1.2 is a straightforward consequence of his results. Set

$$SILT_\epsilon(B) = \int_B dsdt \int_{\Re^d \times \Re^d} f_\epsilon(x-y) X_s(dx)X_t(dy). \tag{1.15}$$

1.2 THEOREM. *Let X_t be a super Brownian motion or super stable process, and let $SILT_\epsilon(B)$ be the approximate intersection local time (1.15). If $d \leq 7$ in the Brownian case, or $d < 4\alpha$ in the stable case, then for all $T \in (0,\infty)$ and all Borel B for which $\bar{B} \cap D = \emptyset$, $L_\epsilon(B)$ converges in $\mathcal{L}^2$ to a finite limit $SILT(B)$ as $\epsilon \to 0$, which we call the SILT of X_t, and which is independent of f.*

In the spirit of (1.13), Dynkin (1988c) obtains the following representation for the SILT in this case:

$$\begin{aligned} SILT(B) &= \int_{s_1,s_2\in B}\int_{t_1,t_2\in B}\int_{\Re^{3d}} p^{\alpha}_{t_1-s_1}(z_1,x)p^{\alpha}_{t_2-s_2}(z_2,x) \\ &\qquad \times Z(ds_1,dz_1)Z(ds_2,dz_2)dx\,dt_1dt_2, \end{aligned} \tag{1.16}$$

where this formula, as do many others to follow, relies heavily on the convention that $p^{\alpha}_t \equiv 0$ for all $t < 0$. A representation in the spirit of (1.12) is the following, in which $\lambda > 0$ and, for simplicity, we take $B = [t_0, t_1] \times [t_2, t_3]$, $t_0 < t_1 < t_2 < t_3$:

$$\begin{aligned} SILT(B) &= \int_{t_0}^{t_1}\int_{t_2}^{t_3} \langle G^{\lambda}_{\alpha}(x-y), X_{s_2}(dx)\rangle\, Z(ds_1,dy)ds_2 \\ &\quad + \lambda\int_{t_0}^{t_1}\int_{t_2}^{t_3} \langle G^{\lambda}_{\alpha}(x-y), X_{s_1}(dx)X_{s_2}(dy)\rangle\, ds_1ds_2 \\ &\quad - \int_{t_2}^{t_3} ds_2 \int_{\Re^d} \langle G^{\lambda}_{\alpha}(x-y), X_{s_2}(dy)\rangle\, (X_{t_1}-X_{t_0})(dx). \end{aligned} \tag{1.17}$$

We shall see how to derive this in Section 5.

The case of $B \cap D \neq \emptyset$ is more delicate, and is best broken into two distinct cases. The first is the case in which a local time exists. Here we have the following result, a more or less immediate consequence of the results of Dynkin (1988c), given that one is prepared to invest the non-trivial effort required in the current case to check that his conditions hold under the conditions of the Theorem.

1.3 THEOREM. *Let X_t be a super Brownian motion or super stable process. If $d \leq 3$ in the Brownian case, or $\alpha > 1 \vee d/2$ in the stable case, then the following limit exists in $\mathcal{L}^2$ for all $T \in (0,\infty)$ and all $\phi \in \mathcal{S}_d$:*

$$\lim_{\epsilon\to 0}\int_0^T dt\int_0^t ds\langle \phi(x)f_{\epsilon}(x-y),\ X_s(dx)X_t(dy)\rangle.$$

The limit, which is independent of f, is called the SILT of X, and can be written as

$$\int_0^T\int_{\Re^d} L^x_t\phi(x)\,X_t(dx)\,dt = \int_0^T \langle L^{\cdot}_t\phi(\cdot), X_t\rangle\,dt,$$

where L^x_t is the local time of Theorem 1.1.

The second case is the truly delicate one, and requires a renormalisation argument, since any attempt at a direct approach to (1.14) as a limit of $L_{\epsilon}(B)$ yields a plethora of infinities. To set up the renormalisation, set

$G_\epsilon^{\alpha\lambda} = G_\alpha^\lambda * f_\epsilon$, where "$*$" denotes convolution, and define a new approximate, renormalised SILT by setting, for every $\phi \in \mathcal{S}_d$,

$$\gamma_\epsilon^\lambda(T,\phi) = \int_0^T dt \int_0^t ds \langle \phi(x) f_\epsilon(x-y),\ X_s(dx)\, X_t(dy)\rangle$$
$$- \int_0^T \langle \phi(x)\, G_\epsilon^{\alpha\lambda}(x-y),\ X_t(dx)\, X_t(dy)\rangle dt. \tag{1.18}$$

Then the following result is due to Adler and Lewin (1992):

1.4 THEOREM. *Let X_t be a super Brownian motion or super stable process, and let $\gamma_\epsilon^\lambda(T,\phi)$ be its approximate renormalised intersection local time. If $d = 4$ or 5 in the Brownian case, or $d/3 < \alpha \le d/2$ in the stable case, then for all $\lambda > 0$, all $T \in (0,\infty)$ and all $\phi \in \mathcal{S}_d$, $\gamma_\epsilon^\lambda(T,\phi)$ converges in $\mathcal{L}^2$ to a finite limit $\gamma^\lambda(T,\phi)$ as $\epsilon \to 0$ which we call the renormalised SILT of X_t, and which is independent of f. Furthermore, γ^λ satisfies the following evolution equation for each $\lambda > 0$:*

$$\gamma^\lambda(T,\phi) = \int_0^T \int_{\Re^d} \left\{ \int_0^t \langle G_\alpha^\lambda(x-y)\phi(x),\ X_s(dx)\rangle\, ds \right\} Z(dt,dy)$$
$$+ \lambda \int_0^T dt \int_0^t ds\, \langle G_\alpha^\lambda(x-y)\phi(x),\ X_s(dx)\, X_t(dy)\rangle \tag{1.19}$$
$$- \int_0^T \langle G_\alpha^\lambda(x-y)\phi(x),\ X_t(dx)\, X_T(dy)\rangle\, dt.$$

Note that while it is tempting to use the simplest form of this result, for which $\lambda = 0$ and there are only two terms on the right hand side of (1.19), there will be occasions (when either G^0 is not defined or we need extra integrability) in which we shall need λ strictly positive.

Note also that there are a number of differences between the SILT of Theorems 1.2 and 1.4. The SILT of the latter involves the "spatial parameter" ϕ, which is not present in the former. Setting $\phi \equiv 1$ would give an SILT in the spirit of Theorem 1.2. The extra parameter has the advantage of indicating "where" intersections are occuring, and is important in setting up interaction mechanisms for more sophisticated particle systems. This difference is actually superficial, since Theorem 1.2 could easily have been rephrased in the more general setting. (Simply add a $\phi(x)$ to the integrals of (1.15) and (1.16).) We have left it as it is for historical consistency only. It *is* important, however, to note that whereas the local time of Theorem 1.1 is point indexed, this is not true for any of our ILT's or SILT's. In fact, it is possible to set up a point indexed version of the SILT γ^λ only in the case $d = 1$.

Furthermore, note that γ^λ is not so much *a* SILT as a *family* of SILT's indexed by the parameter λ. This emphasizes the non-uniqueness of the renormalisation.

Rosen (1990b) has further results on superprocess SILT, involving a somewhat different renormalisation to that above. He also describes what happens to his version of γ_ϵ^λ in the non-convergent case. Note that our renormalisation involves a "doubling" of the measure X – i.e. it involves $X_s \times X_t$ only for $t = s$. It is highly reminiscent of the renormalised, self-intersection local time for $\Re^d$ valued processes, which, as we shall see below, involves the removal of "local double points" of the process.

Before we leave superprocesses, it is interesting to try to understand the various conditions relating α and d on a heuristic basis. A series of results, starting with Dawson and Hochberg (1979) and including Dawson, Iscoe and Perkins (1989) and Perkins (1990) describe in considerable detail the fine, dimensional, structure of the support of superprocesses.

For example, in all of the cases of interest to us, the mass of the random measure $X_t(\cdot)$ is, at any given time t, carried by a (random) set of Hausdorff dimension $d \wedge \alpha$. ($\alpha = 2$ describes the Brownian case.) By this we mean that for each $t > 0$ there exists a random set $\Lambda_t(\omega)$ such that

$$X_t(A \cap \Lambda_t(\omega))(\omega) \;=\; X_t(A)(\omega),$$

for each compact $A \subset \Re^d$ and almost every ω, and that

$$\dim(\Lambda_t(\omega)) \;=\; d \wedge \alpha \qquad \text{for every } \omega.$$

This result can also be made uniform in t, (Perkins (1988)).

Define the "range" of X by

$$R \;=\; \lim_{\epsilon \downarrow 0} \left\{ \text{closure}\Big(\bigcup_{t>\epsilon} S(X_t) \Big) \right\},$$

where $S(\nu)$ denotes the closed support of a measure ν. Then, in the Brownian case, it is also known that the dimension of the range is $4 \wedge d$. The dimensional requirements of Theorems 1.1 and 1.2, at least for the Brownian case, now become more or less obvious. Since the range has dimension $4 \wedge d$ one would expect, in analogy to what is known, for example, of Brownian motion, that the range will include fixed points for dimensions 1,2 and 3. (The case $d = 4$ is not clear from mere dimension calculations – one also needs precise measure function information.)

Using these facts, and counting on one's fingers (also known as "doing Physics") leads one to expect the existence of a local time in these dimensions. Theorem 1.1 says that this is in fact the case. The fact that two $(4 \wedge d)$-dimensional parts of the range will probably intersect in a d-dimensional space if $d \leq 7$ is what gives the dimensional requirements of Theorem 1.2 for the Brownian case. The renormalised self intersection local time of Theorem 1.4 is more subtle, and not amenable to this kind of argument.

The stable case is more complicated, since it is known there that, for each t, $S(X_t)$ is either $\Re^d$ or empty. Nevertheless, the comments made above, together with the results of Theorems 1.1 and 1.2, seem to indicate that the total charge of the measure valued process is carried, in the sense described above, by a set of dimension 2α. There does not seem to be a result of this kind in the literature, but something like it must be true.

This concludes, for the moment, our discussion of superprocess local times.

(c) Local and intersection local times for density processes. The situation for density processes is somewhat different to that for superprocesses, since here the process is a distribution rather than a measure. In this case, we would like to use the following expression to define a new real valued stochastic process, which would serve as the local time process for this case:

$$L_t^x = \int_0^t \eta_s(\delta(\cdot - x))\, ds.$$

Following the same sort of approximations to the delta function as above, one can show that this is well defined only when $d = 1$ and $\alpha > 1$. We shall return to this fact below.

As far as the intersection local time is concerned, it is natural to try to start with the $\mathcal{S}'$ valued process defined by

$$\int_0^t du \int_0^t dv \int_{\Re^d} \int_{\Re^d} (\eta_u \times \eta_v)(\delta(x-y))\, \phi(x)\, dx dy, \tag{1.20}$$

where $\eta_u \times \eta_v$ is the usual product of distributions.

When $d = 1$ and $\alpha > 1$ this is quite simple, for then one can use the basic local time to write (1.20) as $\int_{\Re^d} (L_t^x)^2 \phi(x) dx$. In general, however, one cannot make mathematical sense out of (1.20) without introducing a certain renormalisation. For this, unfortunately, we require some notation.

Let G_1, G_2 be a two zero mean, but otherwise completely general, Gaussian variables. We define their *Wick product* as $: G_1 G_2 := G_1 G_2 - E G_1 G_2$. Equipping $\mathcal{S}_d$ with the usual topology, let $\mathcal{A}_{2d}$ be the dense subset of $\mathcal{S}_{2d}$ made up of functions of the form

$$\phi_N(x_1, x_2) = \sum_{i=1}^N \phi_i^{(1)}(x_1) \cdot \phi_i^{(2)}(x_2), \tag{1.21}$$

where $\phi_i^{(j)}(x) \in \mathcal{S}_d$. If η_1, η_2 are Gaussian distributions on $\mathcal{S}_2$ then we define the corresponding Wick product $:\eta_1 \times \eta_2:$ on $\mathcal{S}_{2d}$ by setting

$$(:\eta_1 \times \eta_2:)(\phi_N) = \sum_{i=1}^{N} :\eta_1(\phi_i^{(1)})\,\eta_2(\phi_i^{(2)}):$$

for test functions of the form (1.21) in $\mathcal{A}_{2d}$ and then extending to all of $\mathcal{S}_{2d}$. That this is legitimate is standard fare in the theory of Gaussian distributions. (See, for example, Chapter 6 of Glimm and Jaffe (1987).)

We are now in a position to make sense out of (1.20). For f, f_ϵ as before, $\phi \in \mathcal{S}_d$, and $t \geq 0$ define the approximate SILT's

$$\begin{aligned}\gamma_\epsilon(t,\phi) &= \gamma_\epsilon(f:t,\phi)\\ &= \int_0^t dt_1 \int_0^t dt_2\,(:\eta_{t_1} \times \eta_{t_2}:)\Big(\phi(x_1)f_\epsilon(x_2-x_1)\Big).\end{aligned} \tag{1.22}$$

The following result is from Adler and Rosen (1992), and has antecedents in Adler, Feldman and Lewin (1991):

1.5 THEOREM. *Let η be the density process corresponding to either a Brownian motion or index α symmetric stable process in $\Re^d$. Assume $d < 2\alpha$ in the stable case, $d \leq 3$ in the Brownian case. Then the approximate SILT's $\gamma_\epsilon(f:t,\phi)$ converge in $\mathcal{L}^2$ as $\epsilon \to 0$. The limit random process $\gamma(t,\phi)$ is independent of the function f, and is called the SILT for η_t.*

Furthermore, for all $\lambda > 0$, the limit γ (which itself is independent of λ) satisfies the following evolution equation:

$$\begin{aligned}\gamma(t,\phi) &= 2\lambda \int_0^t du \int_0^u dv\,(:\eta_u \times \eta_v:)(G_\alpha^\lambda(x-y)\phi(y))\\ &\quad - 2\int_0^t du\,(:\eta_t \times \eta_u:)(G_\alpha^\lambda(x-y)\phi(y))\\ &\quad + 2\int_0^t du\,(:\eta_u \times \eta_u:)(G_\alpha^\lambda(x-y)\phi(y))\\ &\quad + 2\sqrt{2}\int_0^t\int_0^s\int_{\Re^d} \eta_u\big(\Delta_{\alpha/2}G_\alpha^\lambda(x-\cdot)\phi(\cdot)\big)\,du\,W(dx,ds).\end{aligned} \tag{1.23}$$

As before, W is the space-time Gaussian white noise with covariance functional (1.6).

In the Brownian case we can replace the last term in (1.23) by

$$2\int_0^t \int_{\Re^d} \Big\langle \int_0^s \nabla\eta_u\big(G_2^\lambda(x-\cdot)\phi(\cdot)\big)\,du,\ W(dx,ds)\Big\rangle,$$

where W is now a vector valued white noise. Further details, in particular what happens when $d \geq 2\alpha$ and γ_ϵ diverges as $\epsilon \to 0$, can be found in Adler and Rosen (1992).

(d) Brownian motion and stable process local and intersection local times. The title of this section would, of course, make a nice title for a rather thick book. We shall barely scratch the surface of this topic.

The basic existence theorems of Brownian and stable local and intersection local times can all be conjectured once one knows some comparatively simple facts about the dimensional properties of the sample paths of these processes.

In particular, results dating back half a century (see Taylor (1986) for the history and for further details) give us the Hausdorff dimension of the range $R(B) := \{B_t : t \in [0,1]\}$ of a Brownian motion B_t as 2, for all $d \geq 2$. (It is clearly 1 if $d = 1$.) Similarly, the range of a stable process is given by $\alpha \wedge d$.

It is reasonably easy to guess from these results that none of the stable processes will hit points in any dimension other than one. Some more careful calculations show that this is also true for Brownian motion.

Doing Physics, as in §(b) above, allows one to guess (correctly) that, for example, Brownian motion will have multiple points of all orders in dimension 2, double but not triple (or higher order) points in dimension 3, and not even double points in dimensions 4 and higher. Similarly, a stable process will have double points, for example, as long as $d < 2\alpha$. The pre-history of these results dates back at least to Dvoretzky, Erdös and Kakutani (1950), and they and their history are also well summarised in Taylor (1986). It is not hard to relate them to the existence of local and intersection local time.

Local time for a Markov process B can be defined via Markov additive functional (e.g. Blumenthal and Getoor (1968)) or (unfortunately not necessarily consistent) martingale (e.g. Revuz and Yor (1991)) considerations. In the spirit of Geman and Horowitz (1980), however, for us it will simply be the density, with respect to Lebesgue measure, of the occupation measure

$$L_t(A) := \int_0^t 1_A\big(B(s)\big)\, ds.$$

That is, the local time L_t^x satisfies $L_t(A) = \int_A L_t(x)\, dx$ for all Borel $A \subset \Re^d$. It is also realisable as the limit

$$\lim_{\epsilon \to 0} \int_0^t f_\epsilon(B_s - x)\, ds. \tag{1.24}$$

The ILT of two processes B^1 and B^2 can be defined via a similar limit, and for any set $B \subset [0,\infty) \times [0,\infty)$ and any $\phi \in C_b$ we can try to define

$$\begin{aligned} L_B(\phi) &= L_B(\phi\colon B^1, B^2) \\ &= \lim_{\epsilon \to 0} \iint_B f_\epsilon(B^1(t) - B^2(s))\, \phi(B^1(t))\, dtds. \end{aligned} \tag{1.25}$$

When this limit exists, we shall call it the ILT for B^1 and B^2. Note that while the ILT is usually defined without the parameter ϕ, the extra parameter will be important when we come to tying together the ILT's of regular processes and the SILT's of super processes.

If $B^1 = B^2$ in (1.25), and B includes part of the diagonal in $[0,\infty) \times [0,\infty)$ the limit does not generally exist, although there often exists a (random or non-random) renormalising functional F_ϵ so that

$$\begin{aligned} \lim_{\epsilon \to 0} \Big\{ &\iint_B f_\epsilon(B(t) - B(s))\, \phi(B(t))\, dtds \\ &- F_\epsilon\big(B, \{B(s), 0 \le s \le t\}\big) \Big\} \end{aligned} \tag{1.26}$$

is well defined. In such a case we call the limit a ***renormalised*** SILT. The renormalisation is generally not unique. Constructions of this kind, for Brownian motion, date back to Varadhan (1969), with the most recent word on the subject due to Bass and Khoshnevisan (1992). Rosen (1986-1990a, 1991), LeGall (1990) and Yor (1985) contain most of the important results in between these two references.

Remaining for the moment at a heuristic level, recall that hitting points is more or less the same as having a local time, and having double points is like having an SILT. Thus, one would expect that neither Brownian motion nor stable processes will have a local time in dimensions other than 1, and that the ILT between two independent processes will exist in dimensions $d = 1, 2, 3$ for Brownian motion and whenever $d < 2\alpha$ in general. It is a little harder to guess what will happen for the renormalised SILT.

Here are some precise results. In the Brownian motion case they also give Tanaka (evolution) equations. Whereas in principle corresponding equations also exist in the stable case they involve rather complicated integrals against Poisson measures, and do not seem to be very useful. They have not been studied in any detail. Note that this is quite different to the superprocess situation, where Tanaka formulae were, up to the specific form of the infinitesimal generator and Green's function, identical for the Brownian and stable cases.

1.6 Theorem. *Let B_t, $t \ge 0$ be one-dimensional Brownian motion or stable process with $\alpha > 1$. Then the limit in (1.24) exists in $\mathcal{L}^2$, and is called the local time of B. In the Brownian case it satisfies both of the* *evolution equations*

$$L_t^x = |B_t - x| - |B_0 - x| - \int_0^t \mathrm{sgn}(B_s - x)\, dB_s \tag{1.27}$$

$$\tfrac{1}{2}L_t^x = (B_t - x)^+ - (B_0 - x)^+ - \int_0^t 1_{\{B_s > x\}}\, dB_s. \tag{1.28}$$

Formula (1.27) is the original Tanaka formula. (1.28) goes by the same name, as do the "Tanaka formulae" of the preceeding section. For details in the Brownian case my favourite reference is Revuz and Yor (1991). The existence in the stable case is a simple $\mathcal{L}^2$ calculation of the kind that appears in Section 4 for superprocess local times.

Another version of the Tanaka formula, that does not seem to have been used for real valued Brownian motion is the following, which, formally, comes from subsitituting (1.11) into a generalised Itô formula. (cf. Revuz and Yor (1991), Theorem VI.1.1.)

$$\begin{aligned} L_t^x &= -G_2^\lambda(B_t - x) + G_2^\lambda(B_0 - x) \\ &\quad + \int_0^t (G_2^\lambda)'(B_s - x)\, dB_s + \lambda \int_0^t G_2^\lambda(B_s - x)\, ds. \end{aligned} \tag{1.29}$$

The use of the more complicated Green's functions, rather than the relatively simpler functions of Theorem 1.6, comes into its own when treating SILT.

1.7 THEOREM. *Let B^1 and B^2 be independent Brownian motions or stable processes in $\Re^d$. If $d \le 3$ in the Brownian case, or $\alpha > 1 \vee d/2$ in the stable case, then the limit in (1.25) exists in $\mathcal{L}^2$.*

Setting, in the notation of (1.25), $L_t(\phi) = L_{[0,t]\times[0,t]}(\phi)$, we have that in the planar Brownian case L_t satisfies the following evolution equation for every $\lambda > 0$:

$$\begin{aligned} L_t(\phi) &= L_t(\phi : B^1, B^2) \\ &= \lambda \int_0^t du \int_0^u G_2^\lambda(B_u^1 - B_v^2)\phi(B_v^2)\, dv \\ &\quad + \int_0^t G_2^\lambda(B_u^1 - B_u^2)\phi(B_u^2)\, du - \int_0^t G_2^\lambda(B_t^1 - B_u^2)\phi(B_u^2)\, du \\ &\quad + \int_0^t dB_u^1 \int_0^u \nabla G_2^\lambda(B_u^1 - B_v^2)\phi(B_v^2)\, dv. \end{aligned} \tag{1.30}$$

The Tanaka formula here is proven in Adler, Feldman and Lewin (1991), but is really just an easier version of a similar result of Rosen (1986) in which

$B^1 = B^2$. Rosen's result, of course, involves a renormalisation as in (1.26), but since we shall not require this we refer the reader to Rosen's papers for details.

I know of no formula for the SILT of stable processes corresponding to (1.30). Perhaps, given the problems mentioned above in relation to a local time Tanaka formula, this is not surprising.

2. THE PARTICLE PICTURE

In this section we set the background for, and state, our main results. Since these are all related to the particle picture described at the beginning of the previous section, and since we shall require the details of this picture for our proofs, we need to commence with a more careful formulation of the finite system.

If the superprocess limit provides a "bird's eye view" of the large particle system, we require precisely the opposite, described by Ed Perkins as in the heading of the following subsection. The formulation that we shall use is due initially to Walsh (1986), but is basicallly that of Perkins (1990), to whom we refer the interested reader for the finer details of the existence of the setup.

(a) The worm's eye view. We shall start with the branching system of §1(a); i.e. the system in which $\rho < \infty$, and the limit is a measure valued process. Afterwards we shall look at the $\rho = \infty$ situation separately. Furthermore, to save on notation we shall consider only the (typical) case $\rho = 1$. Recall that m is a finite measure of total mass $|m| = m(\Re^d)$.

Let $\{B_t\}_{t\geq 0}$, $B_0 \equiv 0$, be either a Brownian motion with infinitesimal generator Δ, or a symmetric stable process with generator Δ_α. We append to the state space $\Re^d$ a cemetery state Λ, and adopt the convention that $\phi(\Lambda) \equiv 0$ for all functions $\phi : \Re^d \to \Re^k$. Let $\mu > 0$ be a branching rate, and let $K = [\mu|m|]$ be the number of particles alive at time zero, where, as usual, [ ] denotes "integer part".

In order to label our particles, define the family of multi-indices

$$I := \{\alpha = (\alpha_0, \alpha_1 \dots, \alpha_N) \colon \alpha_0 \in \mathcal{N},\ \alpha_i = 1, 2,\ i \geq 1,\ N \in \mathcal{Z}_+\}.$$

Define the "length" of α by $|\alpha| = N$, and set $\alpha|_i = (\alpha_0, \dots, \alpha_i)$ and $\alpha - i = (\alpha_0, \dots, \alpha_{|\alpha|-i})$. Unfortunately α will have a double meaning throughout the remainder of this paper, since, as you will undoubtedly recall, we have already used it to designate the order of a stable process. Nevertheless, it should always be clear from the context which meaning we are using.

We induce a partial order on I by setting

$$\beta \prec \alpha \iff \beta = \alpha|_i \text{ for some } i \leq |\alpha|,$$

and for any $t > 0$ write $\alpha \sim t$, if and only if

$$\frac{|\alpha|}{\mu} \leq t < \frac{1+|\alpha|}{\mu}.$$

For each $\alpha \in I$, let B_t^α be an independent copy of B on the interval $[0, 1/\mu)$, and let N^α be an independent copy of a random variable taking the values 0 and 2 with equal probabilities.

We can now start building the particle system. Choose K points, $z_1, \ldots, z_K$ at random according to the probability measure $m(\cdot)/|m|$. Define a sequence of processes $Y^1, \ldots, Y^K$ on the interval $[0, 1/\mu)$ by setting $Y^i(t) = z_i + B^i(t)$.

For times $t \geq 1/\mu$ we define a binary tree of processes recursively, by setting, for each α,

$$Y^\alpha(t) = \begin{cases} \lim_{t \uparrow \frac{|\alpha|}{\mu}} Y^{\alpha-1}(t) + B^\alpha\left(t - \frac{|\alpha|}{\mu}\right) & \text{if } t \sim \alpha \\ \Lambda & \text{otherwise} \end{cases}$$

The limit here is, of course, from the left. Note that whereas Y_t^α is defined for all t, the process is only "alive" when $t \sim \alpha$.

We are not quite finished, for we have yet to introduce the branching, which we do by thinning out the full tree. For each $\alpha \in I$ define the stopping time

$$\tau^\alpha = \begin{cases} 0 & \text{if } \alpha_0 > K \\ \min_{0 \leq i \leq |\alpha|} \left\{ \frac{i+1}{\mu} : N^{\alpha|i} = 0 \right\} & \text{if this set} \neq \emptyset \text{ and } \alpha_0 \leq K \\ \frac{1+|\alpha|}{\mu} & \text{otherwise} \end{cases}$$

We can now finally define the particle paths of interest to us by setting, for each $\alpha \in I$,

$$X^\alpha(t) = \begin{cases} Y^\alpha(t) & \text{if } t < \tau^\alpha \\ \Lambda & \text{if } t \geq \tau^\alpha \end{cases}$$

As was the case for Y^α, X^α is "alive" for times $t \sim \alpha$. However, there is now also the added restriction that it has a continuous stream of ancestors.

The measure valued process for the finite system is now definable as

$$X_t^\mu(A) = \frac{\#\{X_t^\alpha \in A : \alpha \sim t\}}{\mu}. \tag{2.1}$$

Throughout we shall let $\mathcal{F}_t^\mu$, for $t > 0$, denote the σ-algebra generated by the initial points $z_1, \ldots, z_K$, and all the N^α and $\{Y_s^\alpha,\ s \leq t\}$ with $|\alpha| \leq [t\mu]$.

The basic limit theorem for X^μ is the following, due, initially, to Watanabe (1968):

2.1 THEOREM. *Let $M_F(\Re^d)$ be the space of finite measures on the Borel subsets of $\Re^d$, endowed with the topology of weak convergence. Then, as $\mu \to \infty$, X^μ converges weakly on the Skorohod space $D([0,\infty), M_F(\Re^d))$, and the limit process is the unique solution of the martingale problem (1.2).*

There are a number of ways to prove this result. It is going to be important for us to understand one of them. To this end, we specialise for the moment to the Brownian case, and define two "martingale measures" (cf. Walsh (1986)) as follows:

(2.2)
$$W^\mu((0,t]\times A) = \frac{1}{\sqrt{\mu}}\sum_\alpha \int_0^t 1_A(X_s^\alpha)\, dY_s^\alpha,$$

(2.3)
$$Z^\mu((0,t]\times A) = \frac{1}{\mu}\sum_\alpha 1_A\Big(X^\alpha\Big(\frac{|\alpha|+1}{\mu}\Big)\Big) 1_{[0,\tau^\alpha)}(t)\,(N^\alpha - 1).$$

Here A is a Borel set in $\Re^d$, and you should remember the conventions about when the X^α are alive and dead. The filtration for both martingale measures is $\mathcal{F}_t^\mu$, and the normalisations in μ are chosen so that both W^μ and Z^μ will have well defined weak limits as $\mu \to \infty$.

Fix $t > 0$, and take $\alpha \sim t$. Then Itô's formula states that for any bounded, twice differentiable $\phi : \Re^d \to \Re$,

$$\phi(X_t^\alpha) = \phi(X^\alpha(|\alpha|/\mu)) + \int_{\frac{|\alpha|}{\mu}}^t \nabla\phi(X_s^\alpha)\cdot dY_s^\alpha + \int_{\frac{|\alpha|}{\mu}}^t \Delta\phi(X_s^\alpha)\, ds$$

(2.4)
$$= \phi(X^\alpha(|\alpha|/\mu)) + \int_0^t \nabla\phi(X_s^\alpha)\cdot dY_s^\alpha + \int_0^t \Delta\phi(X_s^\alpha)\, ds,$$

where the change of lower bound on the integral comes from the conventions for $\phi(\Lambda)$.

Summing this over α, while developing a telescopic sum for the first term on the right hand side in (2.4), and keeping in mind the convention about when the X^α are alive, we obtain

$$X_t^\mu(\phi) = X_0^\mu(\phi) + \frac{1}{\mu}\sum_\alpha \phi(X_{\frac{|\alpha|}{\mu}}^\alpha)\,(N^\alpha - 1) + \frac{1}{\mu}\sum_\alpha \int_0^t \nabla\phi(X_s^\alpha)\cdot dY_s^\alpha$$

(2.5)
$$+ \frac{1}{\mu}\sum_\alpha \int_0^t \Delta\phi(X_s^\alpha)\, ds.$$

Rewriting* this in terms of the martingale measures W^μ and Z^μ we obtain

$$\begin{aligned} X_t^\mu(\phi) &= X_0^\mu(\phi) + \int_0^t X_s^\mu(\Delta\phi)\,ds + \int_0^t \int_{\Re^d} \phi(x)\,Z^\mu(ds,dx) \\ &\quad + \frac{1}{\sqrt{\mu}} \int_0^t \int_{\Re^d} \nabla\phi(x)\cdot W^\mu(ds,dx). \end{aligned} \tag{2.6}$$

If we compare (2.6) to the formulation of the limit of X^μ in (1.2a), viz.

$$X_t(\phi) = m(\phi) + \int_0^t X_s(\Delta\phi)\,ds + Z_t(\phi),$$

then it is clear that not only do we have convergence of X^μ to the limit measure diffusion X, but each term in the SPDE seems to converge as well. There is, however, one term that disappears, that involving the stochastic integral with respect to the martingale measure W^μ. While this is hardly surprising, given the extra normalisation of $1/\sqrt{\mu}$, it is precisely the rate of decay of this term that will be the crucial factor determining the results of §2(c) below. Now, however, we shall state a result that formalises the termwise convergence just discussed. Because of the continuous mapping theorem, there are only two terms that need to be considered. The following result can be gleaned from Walsh (1986). To formulate it, we need to treat the martingale measures $Z^\mu(dx,ds)$ and $Z(dx,ds)$ as random distributions, and write, for example, $Z_t^\mu(\phi)$ for $\int_0^t \int_{\Re^d} \phi(x) Z^\mu(dx,ds)$.

2.2 THEOREM. *As $\mu \to \infty$, the vector valued process $\langle X^\mu, Z^\mu \rangle$ converges weakly on the Skorohod space $D([0,\infty), M_F(\Re^d)\times S_d')$, and the limit process $\langle X, Z\rangle$ satisfies (1.2).*

(b) The dimension gap and our main results. In view of Theorem 2.2, so that each term in (2.6) converges weakly to the right limit, it is natural to try to express both local and intersection local time of superprocesses via those of the processes in the underlying particle system. For example, writing $L_t^x(X^\alpha)$ for the local time of X^α, one would like to know whether or not $\mu^{-1}\sum_\alpha L_t^x(X^\alpha)$ converges, as $\mu \to \infty$ to the local time L_t^x of the superprocess.

When $d = 1$ it is not too hard to show that this is the case, and that the following result holds:

*(2.5) and (2.6) are the two most crucial equations in this paper, and, in fact, lie at the core of any understanding of superprocesses in general and the SPDE (1.3) in particular. If you have not seen them before, you should work through them carefully now. If you have troubles, Chapter 8 of Walsh (1986) is a good place to turn to for guidance.

2.3 THEOREM. *We assume the usual branching system of real valued Brownian motions or stable processes with $\alpha > 1$. Then, with the above notation,*

$$\frac{1}{\mu}\sum_{\alpha} L_t^x(X^\alpha) \Rightarrow L_t^x, \qquad \text{as } \mu \to \infty, \tag{2.7}$$

for each fixed $x \in \Re$ and $t > 0$.

Note that this result is not an immediate consequence of Theorem 2.1 and the continuous mapping theorem because of the singularities in the definition of the local times. Nevertheless, it is also not hard. The proof will be given in §4.

It would, of course, be nice to raise the status of (2.7) to a statement about weak convergence in $D([0,1], C(\Re))$. While, in principle, such a result should hold, we shall not attempt to prove it in this paper. We rather hope that it will ultimately be a consequence of some more general theorems that will be the subject of a future paper. When we get to the proof of Theorem 2.3 we shall indicate why, without further work, we cannot prove the more general result.

When $d = 2$ or 3, one cannot even begin to formulate an analogue to Theorem 2.3, since in these cases the local times of the individual Brownian motions are not defined. However, the local time of the superprocess *is* defined. It is this phenomenon that we refer to as the "dimension gap". Despite this problem, one would still like to set up an approximation of some kind.

The way to do this is to work not with the local times of the individual Brownian motions, but a sort of "smeared" local time, given by $\int_0^t f_\epsilon(X_s^\alpha - x)\, ds$ which we know converges to the true local time as $\epsilon \to 0$ when the latter exists. Consider therefore the following "averaged, smeared local time", in which, in order to keep track of the summation, note that implicit in the convention that $\phi(\Lambda) \equiv 0$ for all ϕ is the convention that $\phi(\Lambda - x) \equiv 0$ for all ϕ and x:

$$L_t^{\mu,\epsilon}(x) := \frac{1}{\mu}\sum_{\alpha}\int_0^t f_\epsilon(X_s^\alpha - x)\, ds. \tag{2.8}$$

Note that when $d = 1$ there is no problem in letting $\epsilon \to 0$ for fixed μ, for we simply return to the setup of Theorem 2.3. When $d > 1$, however, the $\epsilon \to 0$ limit, for fixed μ, diverges. In general, therefore, we have to be careful about how we approach the various limits. The next result tells us how to do this, but only for the Brownian case. Recall that m is finite.

2.4 THEOREM. *We assume that the initial measure m has a bounded density with respect to Lebesgue measure. Set $\epsilon = \mu^{-\beta}$, where $0 < \beta < 1/d$ and $d = 2, 3$. Let $L_t^{\mu,\epsilon}$ be the smeared local time of a finite system of branching Brownian particles described above, and L_t the local time of the* *limiting super Brownian motion. Then, for each $x \in \Re^d$, $d = 2, 3$, and $t > 0$,*

$$L_t^{\mu,\epsilon}(x) \Rightarrow L_t^x, \qquad \text{as } \mu \to \infty.$$

The same kind of dimension gap also exists for ILT. Whereas independent Brownian motions have an ILT only up to dimension 3, the super Brownian motion has a SILT up to dimension 5. To handle the lower dimensional cases, set

$$\gamma^\mu(T,\phi) := \frac{1}{\mu^2}\sum_\alpha \sum_{\beta\neq\alpha} L_T(\phi\colon X^\alpha, X^\beta), \tag{2.9}$$

where L_T is the ILT of Theorem 1.7. Note that this average ILT has no need of renormalisation, since the summation is carefully taken over $\alpha \neq \beta$. In fact, we could extend the summation to $\alpha = \beta$, as long as we were prepared to work with the renormalised SILT of the individual particles, a necessary prerequisite since otherwise the individual summands in (2.9) would not exist. However, there are only $O(\mu)$ such terms, and the normalisation in (2.9) is $O(\mu^2)$. Thus, in the limit, these terms would disappear, and not affect the superprocess. Hence we save ourselves some trouble by ignoring them from the very beginning.

2.5 THEOREM. *We assume the usual branching system of Brownian motions in $\Re^d$, $d \leq 3$, or stable processes with $\alpha > 1 \vee d/2$. Then, with the above notation, $\gamma^\mu(T,\phi)$ converges weakly to the superprocess ILT of Theorem 1.3 for each $\phi \in \mathcal{S}$ and $T > 0$.*

Again, it would be nice to have Theorem 2.5 as a richer weak convergence result, this time in $D([0,1], \mathcal{S}'(\Re^d))$. Once again, however, this will have to wait for a future paper.

The higher dimensional cases are somewhat more delicate, and, as before, we specialise to the Brownian case. Despite the fact that we shall continue to avoid the self intersection local times of the individual particles, there is nevertheless a need for renormalisation. As before, we choose a $\lambda > 0$ with which to do this. Set

$$\begin{aligned}\gamma^{\mu,\epsilon}(T,\phi) := &\frac{1}{\mu^2}\sum_\alpha\sum_{\beta\neq\alpha}\int_0^T dt \int_0^t f_\epsilon(X_s^\alpha - X_t^\beta)\,\phi(X_t^\beta)\,ds \\ &- \frac{1}{\mu^2}\sum_\alpha\sum_{\beta\neq\alpha}\int_0^T (G^\lambda * f_\epsilon)(X_t^\alpha - X_t^\beta)\phi(X_t^\alpha)\,dt.\end{aligned} \tag{2.10}$$

where G^λ denotes the Green's function of Brownian motion. We have

2.6 THEOREM. *Assume that the initial measure has a bounded density with respect to Lebesgue measure. Set* $\epsilon = \mu^{-\beta}$, *where* $0 < \beta < 1/(d-2)$ *and* $d = 4, 5$. *Let* $\gamma^{\mu,\epsilon}(T,\phi)$ *be the smeared intersection local time of a finite system of branching Brownian particles, as described above, and* $\gamma(T,\phi)$ *the intersection local time of the limiting super Brownian motion. Then, with* $d = 4, 5$, *we have that for each* $\lambda > 0$, $T > 0$ *and* $\phi \in \mathcal{S}$,

$$\gamma^{\mu,\epsilon}(T,\phi) \Rightarrow \gamma(T,\phi), \qquad \text{as } \mu \to \infty.$$

Theorems 2.4 and 2.6 represent the main results of this paper. It is unfortunate that we do not have analogues of either of them for the stable case. This deficiency is a consequence of our method of proof, which relies heavily on Itô's formula and the related Tanaka formulae for local and intersection local times of the individual particle motions. Recall that in the stable case we do not have a convenient Tanaka formula for the local time of the individual stable motions, although we do have a Tanaka formula for the local time and SILT of the stable superprocesses.

Nevertheless, similar results for the stable case must hold, and it is not hard, from the calculations with Green's functions in §4, to guess what the correct scaling between ϵ and μ must be.

Our method of proof does extend quite straightforwardly to the case in which the individual particles are elliptic diffusions, but I have not carried out any of the details of the computations for this case.

Before we turn to proofs, let us look at a simpler situation, in which there is no dimension gap.

(c) The Brownian density process. Since the primary aim of this section is to show that there is no dimension gap in the absence of branching, we can save considerable time and space if we change the setup a little so that we can use known results. Again, we restrict attention to the Brownian case.

Rather than define the Brownian density process as the limit of the non-branching ($\rho = \infty$) system normalised as in (1.4), consider the following setup:

Let $X_t^1, X_t^2, \ldots,\ t \geq 0$ be a sequence of independent, $\Re^d$ valued Brownian motions with generator Δ, and with initial values given by the points of a Poisson point process on $\Re^d$ of intensity $\mu > 0$. Let $\sigma^1, \sigma^2, \ldots$ denote a sequence of independent Rademacher random variables. ($P\{\sigma^i = +1\} = P\{\sigma^i = -1\} = \frac{1}{2}$). All random elements are assumed independent of one another except for the fact that the Poisson process determines the initial values of the X^i.

For $\phi \in \mathcal{S}_d$, let ν_t^μ be the $\mathcal{S}_d'$ valued random process defined by

$$\nu_t^\mu(\phi) = \frac{1}{\sqrt{\mu}} \sum_{i=1}^{\infty} \sigma^i \phi(X_t^i). \tag{2.11}$$

Whereas this is a somewhat different process to the η_t^μ of (1.4), it turns out that it has the same limit as $\mu \to \infty$, in that its limit solves the same SPDE, albeit with a slightly different initial condition. The local "sample path" properties of the two processes are identical, and so we can study either of the two particle pictures in trying to understand the limit.

With $L_t^x(X^i)$ the local time of X^i, and $L_t(\phi: X^i, X^j)$ the usual ILT of X^i and X^j, set

$$L_t^\mu(x) = \frac{1}{\sqrt{\mu}} \sum_i \sigma^i L_t^x(X^i), \tag{2.12}$$

and

$$\gamma^\mu(t,\phi) = \frac{1}{\mu} \sum_{i \neq j} \sigma^i \sigma^j L_t(\phi; X^i, X^j). \tag{2.13}$$

Note that as is the case for all the processes in the density case, both L_t^μ and γ^μ are signed processes.

The SILT part of the following theorem appears explicitly in Adler, Feldman and Lewin (1991) (modulo some minor differences in constants) while the local time part can be proven by refining the results of Adler and Epstein ($\equiv$ Feldman) (1987) in the spirit of the proofs in the former paper. We leave the details to the reader.

2.7 THEOREM. *If $d = 1$, then for each x and t*

$$L_t^\mu(x) \Rightarrow L_t(x), \qquad \text{as } \mu \to \infty,$$

where $L_t(x)$ is the local time of the Brownian density process.

If $d \leq 3$ and $\phi \in \mathcal{S}$ then

$$\gamma^\mu(t,\phi) \Rightarrow \gamma(t,\phi), \qquad \text{as } \mu \to \infty.$$

What these results show is that there is an approximation via a particle system for both the local time and SILT of density processes *whenever these exist.* Since we have already seen that this cannot be the case for superprocesses, the meaning of the combined results is clear: The dimension gap is not generated by the increased intensity of particles, but is a consequence of the well known clumping phenomena associated with the branching process.

3. MOMENT FORMULAE

An important component in virtually all the proofs of the following section will be the evaluation of moments of the kind $E \prod_{i=1}^k \langle \phi_i, X_{t_i}^\mu \rangle$. An algorithm for calculating such moments, but for the limit superprocesses, and based exclusively on the transition densities $p_t(x,y)$ and $p_t^\alpha(x,y)$ of Section 1, has been given by Dynkin (1988c). For comparison purposes, and for ease of exposition, we first state the results of Dynkin. Doing so will also enable us to simplify some of the proofs of the following subsection. Dynkin's formulae are, as is to be expected, neater and more compact than those for the particle system.

The main results of this section are the moment formulae for finite particle systems given in §3(b). In principle, these should be "well known", and are really a natural way to arrive at the Dynkin formulae. Nevertheless, they do not seem to be written down anywhere.

(a) Dynkin's diagrams for superprocesses. Since we lose no generality by doing so, for the remainder of this section we shall take m absolutely continuous with respect to Lebesgue measure, but not necessarily finite, and test functions $\phi_i \in \mathcal{L}^p(m)$, for some $p > 1$.

Dynkin's original formulation of the following result, which covered a much wider variety of superprocesses than we are concerned with, was for positive, measurable test functions ϕ and finite initial measures m. That the result is true for initial measures and test functions of the type just described is a consequence of Dynkin's proof and Theorem 2.3 of Adler and Lewin (1992).

3.1 THEOREM. *Let X be a super Brownian motion or super stable process. Then, for all $k \geq 1$ and $\phi_1, \ldots, \phi_k$ as described above,*

$$E\Big(\prod_{i=1}^{k}\langle \phi_i, X_{t_i}\rangle\Big) = \sum_{D_k} 2^{-|V_o|}\int \prod_{v\in V_-} m(dx_v)\prod_{\ell\in L} p^{\alpha}_{s_{f(\ell)}-s_{i(\ell)}}(y_{f(\ell)}-y_{i(\ell)})$$

$$\times \prod_{v\in V_o} ds_v dy_v \prod_{i=1}^{k}\phi_i(z_i)dz_i \tag{3.1}$$

where $p_s^\alpha(x) \equiv 0$ if $s < 0$, and D_k is the set of directed binary graphs with k exits marked $1, 2, \ldots, k$. Given such a graph, L is the set of directed links, and if the link $\ell \in L$ goes from vertex v to vertex w, we write $v = i(\ell)$, $w = f(\ell)$. To each vertex v we associate two variables,

$$(s_v, y_v) \in \Re_+ \times \Re^d,$$

which we refer to as the time and space coordinates of v. V_- denotes the set of entrances for our graph, and if $v \in V_-$ we set $s_v = 0$ and $y_v = x_v$. If v is the exit labelled by j, $1 \leq j \leq k$, we set

$$(s_v, y_v) \equiv (t_j, z_j).$$

Finally, V_o denotes the set of internal vertices, i.e. those vertices that are neither entrances nor exits, and $|V_o|$ is the number of vertices in V_o.

Note that the above formulae differ from those appearing in Dynkin's paper by the factor $2^{-|V_o|}$. This is due to the fact that Dynkin uses a slightly different normalisation in his particle picture, and so considers a slightly different superprocess to those considered here.

(b) Moment diagrams for the particle picture. We shall present three separate results in this subsection. The first is an easy, explicit formula for first moment of $X_t^\mu(\phi)$. The second is a detailed expression for second, mixed moments. This is important, for this is a case in which it is easy to write out a full proof. From this we generalise to a general moment formula for all orders.

3.2 THEOREM. *Let $\phi \in \mathcal{L}^p(m)$ for some $p \geq 1$. Then, under the usual setup,*

$$E(\langle \phi, X_t^\mu \rangle) = \left(\frac{K}{\mu|m|}\right) \cdot \int_{\Re^{2d}} p_t^\alpha(x,y)\,\phi(y)\,m(dx)dy. \tag{3.2}$$

(Recall that $K \sim \mu$ is the initial number of particles.)

PROOF: We need to calculate $\mu^{-1}\sum_{\alpha\sim t} E\phi(X_t^\alpha)$. There are $K2^{[\mu t]}$ α's in this sum, and for each one of these X_t^α is "alive" with probability $2^{-[\mu t]}$, since there are $[\mu t]$ generations of binary branching until time t. Since $\phi(X_t^\alpha) \equiv 0$ if X_t^α is not alive, it follows that $E\langle\phi, X_t^\mu\rangle = (K/\mu)E\phi(X_t^\alpha)$. Since the way we have pieced together the paths of $X^{\alpha-|\alpha|}, X^{\alpha-|\alpha|+1}, \ldots, X^\alpha$ gives one long path of the Markov process with transition density $p_t^\alpha(x,y)$ and initial distribution $m/|m|$, (3.2) is now immediate. ∎

3.3 THEOREM. *Let $\phi_1, \phi_2 \in \mathcal{L}^p(m)$, $p \geq 1$, and let $N_1 = [t_1\mu]$, $N_2 = [t_2\mu]$. Then*

$$\begin{aligned}
E(\langle \phi_1, X_{t_1}^\mu\rangle\langle\phi_2, X_{t_2}^\mu\rangle)
&= \left(\frac{K(K-1)}{(\mu|m|)^2}\right)\cdot\left(\int m(dx)\int p_{t_1}^\alpha(x,y)\,\phi_1(y)\,dy\right)\\
&\qquad\qquad \times\left(\int m(dx)\int p_{t_2}^\alpha(x,y)\,\phi_2(y)\,dy\right)\\
&\quad + \left(\frac{\frac{1}{2}K}{\mu^2|m|}\right)\cdot\sum_{k=1}^{N_1\wedge N_2}\int m(dx)\int p_{\frac{k}{\mu}}^\alpha(x,z)\Big(\int p_{t_1-\frac{k}{\mu}}^\alpha(z,y_1)\phi_1(y_1)\,dy_1\\
&\qquad\qquad \times\int p_{t_2-\frac{k}{\mu}}^\alpha(z,y_2)\phi_2(y_2)\,dy_2\Big)\,dz\\
&\quad + \left(\frac{\frac{1}{2}K}{\mu^2|m|}\right)\cdot\iiint p_{t_1}^\alpha(x,y_1)p_{t_2-t_1}^\alpha(y_1,y_2)\,\phi_1(y_1)\phi_2(y_2)\,m(dx)dy_1dy_2,
\end{aligned} \tag{3.3}$$

where all integrals are over $\Re^d$.

Before we prove this result, we note two important aspects of (3.3). The first thing that should be noted is that, since $K = O(\mu)$, the last term above is $O(1/\mu)$ for large μ, and thus, if the integral is finite, will disappear in the limit. After you have read the proof of the Theorem, you will see why this is the case. (The disappearing term corresponds to certain particle configurations that, proportionately, become increasingly rare as the number of particles increases.) The other point is that, in the limit, the summation in the second term turns into an integral. This process absorbs one of the multiplicative factors of $1/\mu$, while the second one cancels with K. It is then easy to check that what remains is precisely the Dynkin diagram formula of Theorem 3.1. Since our proof of Theorem 3.3 is, of necessity, based on particle paths, it follows that we have an interpretation of Dynkin's result for this case, in which the various diagrams will take on an intuitive meaning associated with sample path behaviour. This "interpretation" could, in fact, be built into a proof, albeit with a non-trivial amount of moment estimation.

Proof of Theorem 3.3: To make this and the following proof a little easier, write $I_\alpha(t)$ for the indicator of the event that X^α is alive at time t; i.e.

$$I_\alpha(t) = \begin{cases} 1 & \text{if } \alpha \sim t \text{ and } X_t^\alpha \neq \Lambda \\ 0 & \text{otherwise} \end{cases}$$

We seek

$$\begin{aligned} E\big(\langle\phi_1, X_{t_1}^\mu\rangle\langle\phi_2, X_{t_2}^\mu\rangle\big) &= \frac{1}{\mu^2}\sum_{\alpha\sim t_1}\sum_{\beta\sim t_2} E\{\phi_1(X_{t_1}^\alpha)\phi_2(X_{t_2}^\beta)\, I_\alpha(t_1)I_\beta(t_2)\} \\ &= \frac{1}{\mu^2}\sum_{\alpha\sim t_1}\sum_{\beta\sim t_2} E\{\phi_1(X_{t_1}^\alpha)\phi_2(X_{t_2}^\beta)\big|I_\alpha(t_1)I_\beta(t_2)=1\} \\ &\qquad\times P\{I_\alpha(t_1)I_\beta(t_2)=1\} \qquad (3.4) \\ &= \frac{1}{\mu^2}\sum_{\alpha\sim t_1}\sum_{\beta\sim t_2} E\{\phi_1(Y_{t_1}^\alpha)\phi_2(Y_{t_2}^\beta)\}\cdot E\{I_\alpha(t_1)I_\beta(t_2)\}, \end{aligned}$$

where the Y^α, defined at the beginning of Section 2, form the full binary tree of processes that we thinned, by demanding a continuous stream of ancestors, to obtain the X^α, and we have used the independence of the basic random motions and the branching.

Assume, with no loss of generality, that $t_1 < t_2$. (The case $t_1 = t_2$ is virtually identical, but simpler, and is left to the reader.) Furthermore, let us use α as it were the "name" of a child on the full binary tree. Then there are three cases to consider:

(i) $\alpha - i \neq \beta - j$ for any $1 < i \leq |\alpha|$ or $1 < j \leq |\beta|$. In this case α and β are the descendents of different ancestors in the initial $t = 0$ generation.
(ii) $\alpha = \beta - i$ for some $1 < i \leq |\beta|$. In this case α is a direct ancestor of β.
(iii) $\alpha - i = \beta - j$ for some $1 < i \leq |\alpha|$ and $1 < j \leq |\beta|$, but $\alpha \not\prec \beta$. In this case α and β have a common ancestor, but α is not a direct ancestor of β. (Since $t_1 < t_2$, β cannot be an ancestor of α.)

CASE (I): Since α and β live on completely separate trees, both expectations at the end of (3.4) split into products of expectations, and the argument used to prove (3.2) gives the first term of (3.3), the factor of $K(K-1)$ coming, obviously, from the number of choices of pairs of zero-th generation parents.

CASE (II): This will give us the third, final term of (3.3). Since α and β live on the same tree, the factor of K in this term comes from the K choices of initial ancestor, while the μ^{-2} comes from the normalisation. Let $0 < k \leq |\beta|$ be such that $|\beta| - k = |\alpha| = N_1$. Then, with an eye on (3.4), and noting that the sum over α there now contains only one term, we have

$$\begin{aligned}
&\sum_{\beta \sim t_2} E\{\phi_1(Y_{t_1}^{\beta-k})\phi_2(Y_{t_2}^{\beta})\} \cdot E\{I_{\beta-k}(t_1)I_\beta(t_2)\} \\
(3.5)\qquad &= \sum_{|\beta|=N_2} E\Big\{\phi_1(Y_{t_1}^{\beta-k})I_{\beta-k}(t_1)E\{\phi_2(Y_{t_2}^{\beta})I_\beta(t_2)\big|\mathcal{F}_{t_1}^{\mu}\}\Big\}.
\end{aligned}$$

The Markov nature of the Y^α and the criticality of the branching give that the conditional expectation is equal to

$$2^{-k}\int \phi_2(y)p_{t_2-t_1}^{\alpha}(Y_{t_1}^{\beta-k}, y)\, dy.$$

Each summand of (3.5) therefore becomes

$$2^{-k}2^{-(N_2-(k+1))}\iiint p_{t_1}^{\alpha}(x, y_1)p_{t_2-t_1}^{\alpha}(y_1, y_2)\phi_1(y_1)\phi_2(y_2)\, m(dx)dy_1dy_2,$$

and summing over β gives the required result. Note that the factor of $\frac{1}{2}$ that arises here is what corresponds to the factor of $2^{-|V_\circ|}$ in (3.1), and in general we shall obtain a factor of $\frac{1}{2}$ at every branch point of this kind.

CASE (III): This is the hardest term, and also the most typical of the higher moment cases. As in Case (ii), both α and β descend from the same (one of K) initial ancestors, which fact, together with the normalisation, gives the multiplicative factor in the middle term of (3.3).

Suppose that the common ancestor of α and β is born at time k/μ, where, of necessity, $k < |\alpha| < |\beta|$. Allowing k to run gives the summation in (3.3), and leaves us with the following summand for each k:

$$\sum_{\alpha}\sum_{\beta} E\{\phi_1(X^{\alpha}_{t_1})\phi_2(X^{\alpha}_{t_2})\} =$$

$$\sum_{\gamma:\,|\gamma|=k}\ \sum_{\alpha}\sum_{\beta} E\Big\{I_\gamma(k/\mu)E\{I_\alpha(t_1)\phi_2(Y^{\alpha}_{t_1})|\mathcal{F}^{\mu}_{\frac{k}{\mu}}\}E\{I_\beta(t_2)\phi_2(Y^{\beta}_{t_2})|\mathcal{F}^{\mu}_{\frac{k}{\mu}}\}\Big\}$$

It is now obvious how to complete the argument and so obtain (3.3). ∎

In fact, a far more general result than Theorem 3.3 also holds:

3.4 THEOREM. *Under the same conditions as in Theorem 3.1, the mixed moments* $E\big(\prod_{i=1}^{k}\langle\phi_i,\, X^{\mu}_{t_i}\rangle\big)$ *are, up to an additional term of* $O(1/\mu)$, *given by the right hand side of (3.1), with two simple changes:*
(a) The time variables s_v *need to be replaced by indices* j_v/μ, *and the corresponding integrals by summations over the integer valued* j_v. *The upper limits on the summations are of the form* $[\mu t_v]$;
(b) Each summation carries with it a multiplicative factor of K/μ^2.

Note that implicit in the above result is the requirement that the branching points (in time) of the binary graphs of Theorem 3.4 are at integer multiples of $1/\mu$.

PROOF: If we start with the case $k = 2$, then all we need to do is note that the last term of (3.3) is $O(1/\mu)$, and we are done. The important, qualitative fact to note at this point is that this was the term that came from pairs (α,β) of particles for which β was a direct descendent of α.

Following the argument used above now for general k, it is easy to see that the graphs described in Theorem 3.1 all correspond to moments involving sets of k particles in which no pair is related via direct descent. There are other terms, in which $1 \le m_2 \le [k/2]$ pairs are so related, or in which $1 \le m_3 \le [k/3]$ triples form parent/child/grandchild type relationships, etc. These terms are, respectively, $O(\mu^{-m_2})$, $O(\mu^{-m_3})$, etc. This way you can prove the theorem in the general case.

If you find the above proof somewhat lacking in detail, you are invited to write out the full argument for the case $k = 4$. This is tedious, but still quite easy. It is also the highest order case that we shall need in this paper.

∎

4. PROOFS OF THE MAIN RESULTS

(a) Local time. We start with something easy:

PROOF OF THEOREM 2.3: We are required to prove the weak convergence of $L^{\mu}_t(x) := \mu^{-1}\sum_{\alpha} L^{x}_t(X^{\alpha})$ to the local time L^{x}_t of the super process, in the case $d = 1$, and start by noting three facts:

Firstly, by Theorem 1.6, the approximate local time

$$L_t^{\alpha,\epsilon}(x) = \int_0^t f_\epsilon(X_s^\alpha - x)\,ds$$

converges, in $\mathcal{L}^2$, as $\epsilon \to 0$, to the local time $L_t^\alpha(x)$ of the process X^α. Consequently, for each fixed μ, the average "smeared local times",

$$L_t^{\mu,\epsilon}(x) = \mu^{-1}\sum_\alpha L_t^{\alpha,\epsilon}(x),$$

converge in $\mathcal{L}^2$, as $\epsilon \to 0$, to the L_t^μ defined above. Secondly, by Theorem 2.1 and the continuous mapping theorem, we know that

$$L_t^{\mu,\epsilon}(x) \Rightarrow L_t^\epsilon(x) := \int_0^t \langle f_\epsilon(\cdot - x), X_s\rangle\,ds$$

as $\mu \to \infty$, for each fixed $\epsilon > 0$. Note for later reference that, at this stage, with x and $\epsilon > 0$ fixed, the convergence can be taken in $D([0,1],\Re)$.

Finally, by Theorem 1.1, $L_t^\epsilon(x) \stackrel{\mathcal{L}^2}{\to} L_t(x)$. Thus, for fixed δ_i, we can now argue as follows:

$$(4.1)\quad P\{L_t^\mu(x) \le a\} \le P\{L_t^\mu(x) \le a;\ |L_t^\mu(x) - L_t^{\mu,\epsilon}(x)| \le \delta_1\} + \eta_1(\epsilon),$$

where the fact that η_1 is independent of μ is a function of the specific form of $L_t^\mu(x) - L_t^{\mu,\epsilon}(x)$ as a sum of iid differences of individual local times, each of which converges in $\mathcal{L}^2$. Continuing, we have

$$\begin{aligned} P\{L_t^\mu(x) \le a\} &\le P\{L_t^{\mu,\epsilon}(x) \le a + \delta_1\} + \eta_1(\epsilon) \\ &\le P\{L_t^\epsilon(x) \le a + \delta_1\} + \eta_1(\epsilon) + \eta_2(\epsilon,\mu), \end{aligned}$$

where $\eta_2(\epsilon,\mu) \to 0$ as $\mu \to \infty$ for any $\epsilon > 0$. Arguing as above,

$$\begin{aligned} &P\{L_t^\mu(x) \le a\} \\ &\quad \le P\{L_t^\epsilon(x) \le a + \delta_1;\ |L_t^\epsilon(x) - L_t(x)| \le \delta_2\} + \eta_1(\epsilon) + \eta_2(\epsilon,\mu) + \eta_3(\epsilon) \\ &\quad \le P\{L_t(x) \le a + \delta_1 + \delta_2\} + \eta_1(\epsilon) + \eta_2(\epsilon,\mu) + \eta_3(\epsilon). \end{aligned}$$

Finally, we have

$$\begin{aligned} \lim_{\mu\to\infty} P\{L_t^\mu(x) \le a\} &\le \lim_{\epsilon\to 0}\lim_{\mu\to\infty} P\{L_t^\mu(x) \le a\} \\ &\le P\{L_t(x) \le a + \delta_1 + \delta_2\}. \end{aligned}$$

A similar argument gives a similar equality with the rôles of L_t^μ and L_t reversed, which suffices to prove the theorem. ∎

Note that because of the convergence in $D([0,1],\Re)$ noted in the proof, if we knew that $\sup_{t\in[0,1]}|L_t^\epsilon(x)-L_t(x)|\rightarrow_{\mathcal{L}^2}0$, then we would be able to lift the status of the entire result to the full weak convergence of the real valued processes $L^\epsilon_\cdot(x)$. The lack of the uniform $\mathcal{L}^2$ convergence is what stops us getting the stronger result. Of course, even more would be needed to get the most general result, involving the weak convergence of L^ϵ as an element of $D([0,1],C(\Re))$.

We now turn to the major task of this subsection – the proof of Theorem 2.4. We start with a Tanaka-like representation for the averaged, smeared, local times, $L_t^{\mu,\epsilon}$. To save on notation, fix $\lambda\geq 0$ and set

$$K_\epsilon(x) \;=\; (G^\lambda * f_\epsilon)(x),$$

where both G^λ and f_ϵ are as defined in §2. (In the more general notation of Section 1, K_ϵ was $G_\epsilon^{\alpha\lambda}$ with $\alpha=2$. Here we are trying to save on subscripts.) Note that K_ϵ is both bounded and infinitely differentiable, and, in view of (1.11), satisfies the equation

$$(-\Delta+\lambda)K_\epsilon \;=\; f_\epsilon. \tag{4.2}$$

Furthermore, $K_\epsilon\rightarrow G^\lambda$, pointwise, as $\epsilon\rightarrow 0$.

Using the smoothness properties of K_ϵ, along with (4.2), we can substitute into (2.6) to immediately obtain

4.1 LEMMA. *Let $L_t^{\mu,\epsilon}$ be the averaged, smeared, local time of a system of branching Brownian motions in $\Re^d$, $d=2,3$. (c.f. (2.8).) Then $L_t^{\mu,\epsilon}$ satisfies the evolution equation*

$$\begin{aligned} L_t^{\mu,\epsilon}(0) \;=\; & -X_t^\mu(K_\epsilon) \;+\; X_0^\mu(K_\epsilon) \;+\; \lambda\int_0^t X_s^\mu(K_\epsilon)\,ds \\ & +\int_0^t\int_{\Re^d} K_\epsilon(x)\,Z^\mu(ds,dx) \\ & -\frac{1}{\sqrt{\mu}}\int_0^t\int_{\Re^d}\nabla K_\epsilon(x)\cdot W^\mu(ds,dx), \end{aligned} \tag{4.3}$$

where the martingale measures W^μ and Z^μ are as defined at (2.2) and (2.3). An equivalent equation holds for $L_t^{\mu,\epsilon}(y)$, $y\in\Re^d$, if K_ϵ is replaced by $K_\epsilon(\cdot-y)$ throughout.

Since the $\epsilon\rightarrow 0$ limits of all but the last term in (4.3) also appear in the Tanaka formula (1.12) for the superprocess local time L_t, it is easy to guess that the main part of the proof of Theorem 2.4 will be in obtaining conditions to ensure that the last term will in fact disappear. The normalisation of $1/\sqrt{\mu}$ makes it seem feasible that this can be done, but the fact that ∇K_ϵ has poor integrability properties, that become worse as $\epsilon\rightarrow 0$, shows that it is going to be necessary to carefully tie together the rates at which $\epsilon \to 0$ and $\mu \to \infty$. Thus it is the following Lemma that is the crux of the proof of Theorem 2.4.

4.2 LEMMA. *Set $\epsilon = \mu^{-\beta}$, where $0 < \beta < 1/d$ for $d = 2, 3$. Assume that the initialising measure m is has a bounded density with respect to Lebesgue measure. Then, for all $\lambda > 0$,*

$$\lim_{\mu\to\infty} E\Big\{\Big(\frac{1}{\sqrt{\mu}}\int_0^t \int_{\Re^d} \nabla K_\epsilon(x)\cdot W^\mu(ds,dx)\Big)^2\Big\} = 0. \tag{4.4}$$

PROOF: The proof is a reasonably straightforward moment calculation, based on Theorem 3.2. (This is despite the fact that Theorem 3.2 is about first moments, and (4.4) is about second moments!) We shall need the fact, easily checkable via a direct calculation from the definition (2.2) of W^μ, that for each bounded, continuous ϕ the martingale $W_t^\mu(\phi)$ (in t) has quadratic variation

$$\langle W^\mu(\phi)\rangle_t = \Big(\int_0^t X_s^\mu(\phi^2)\,ds\Big)\, I, \tag{4.5}$$

where I is the $d \times d$ unit matrix. (It is this fact that reduces the moment calculation by one power.) It therefore follows that

$$\begin{aligned} &E\Big\{\Big(\frac{1}{\sqrt{\mu}}\int_0^t \int_{\Re^d} \nabla K_\epsilon(x)\cdot W^\mu(ds,dx)\Big)^2\Big\} \\ &\qquad\qquad = \frac{1}{\mu}\int_0^t E\{X_s^\mu(\|\nabla K_\epsilon\|^2)\}\,ds. \end{aligned} \tag{4.6}$$

Let T_t, $t \geq 0$, be the semigroup for the Brownian generator Δ. Then, by Theorem 3.2, we have that the right hand side of (4.6) is equivalent to

$$\begin{aligned} \frac{K}{\mu^2|m|}\int_0^t \langle T_s(\|\nabla K_\epsilon\|^2), m\rangle\,ds &\leq \frac{\text{const}}{\mu}\int_0^t \langle T_s(\|\nabla K_\epsilon\|^2), Leb\rangle\,ds \\ &\leq \frac{t\cdot\text{const}}{\mu}\int_{\Re^d} \|\nabla K_\epsilon(x)\|^2\,dx, \end{aligned} \tag{4.7}$$

using firstly the facts that $K \sim \mu$ and m has a bounded density with respect to Lebesgue measure, and then that T_t is a contraction operator on $\mathcal{L}^1$.

In view of (4.7), all that remains is to determine the divergence rate of the $\mathcal{L}^2$ norm of ∇K_ϵ as $\epsilon \to 0$. This is an easy Fourier transform calculation, for which we need to know that $\hat{G}^\lambda(p) = 1/(\lambda + \|p\|^2)$ for all $d \geq 2$, where, as usual, $\hat{}$ denotes a Fourier transform. We have

$$\begin{aligned}\int_{\Re^d} \|\nabla K_\epsilon(x)\|^2\,dx &= \int_{\Re^d} \|p\|^2 \left(\hat{G}^\lambda(p)\right)^2 \hat{f}_\epsilon^2(p)\,dp \\ &= \int_{\Re^d} \|p\|^2 \left(\hat{G}^\lambda(p)\right)^2 \hat{f}^2(p\epsilon)\,dp.\end{aligned}$$

Break the integral up into $\|p\| \le \epsilon$ and $\|p\| > \epsilon$. It is easy to see that the first integral is $O(\epsilon^{d+2})$, and so small. In view of the form of $\hat{G}^\lambda$, and the fact that λ is strictly positive, the second term is bounded by a constant multiple of

$$\int_{\|p\|>\epsilon} \hat{f}^2(p\epsilon)\,dp \;=\; O(\epsilon^{-d}).$$

Substituting into (4.7) gives an overall bound of $O(\mu^{-1}\epsilon^{-d})$, which completes the proof of the Lemma. ∎

PROOF OF THEOREM 2.4: In order to save on notation, we shall treat only the (representative) case $x = 0$. The changes necessary to build a proof for general x are trivial. In view of the evolution equation (4.3) describing $L_t^{\mu,\epsilon}$, and the evolution equation (1.12) describing the superprocess local time, it is clear that with Lemma 4.2 behind us it will suffice to show, for fixed t, the weak convergence

$$\begin{aligned}&\left(X_t^\mu(K_\epsilon),\, \int_0^t\int_{\Re^d} K_\epsilon(x)\, Z^\mu(ds,dx)\right) \\ (4.8)\qquad &\Rightarrow \left(X_t(G^\lambda),\, \int_0^t\int_{\Re^d} G^\lambda(x)\, Z(dx,ds)\right),\end{aligned}$$

as $\mu \to \infty$, where we now set $\epsilon = \mu^{-\beta}$, and β statisfies the conditions of the Theorem. (In fact, as will be obvious from the proof, (4.8) holds even if $\epsilon \to 0$ independently of μ. It is only the term handled by Lemma 4.2 that requires the linkage.) Consider the first of the terms in (4.8). For arbitrary $\eta > 0$,

$$X_t^\mu(K_{\mu^{-\beta}}) \;=\; X_t^\mu(K_{\mu^{-\beta}} - K_\eta) \;+\; X_t^\mu(K_\eta).$$

Assume that we could show that

$$(4.9)\qquad \lim_{\eta\to 0}\lim_{\mu\to\infty} E\left[X_t^\mu(K_{\mu^{-\beta}} - K_\eta)\right]^2 \;=\; 0.$$

Then the facts that for fixed η it is immediate that $X_t^\mu(K_\eta) \Rightarrow X_t(K_\eta)$ as $\mu \to \infty$ (cf. Theorem 2.1) and that $X_t(K_\eta) \to_{\mathcal{L}^2} X_t(G^\lambda)$ as $\eta \to 0$, (intrinsic to the proof of Theorem 1.1; cf. Adler and Lewin (1992)) are enough to finish the proof.

Theorem 3.3 is what we need to handle (4.9), and this gives us that there are three different terms (cf. (3.3)) to calculate. We shall only look at the middlemost, most complicated term of (3.3), and leave the others to the reader. In our case this is

$$\left(\frac{K}{\mu^2|m|}\right)\cdot\sum_{k=1}^{N}\int m(dx)\int dz\, p_{\frac{k}{\mu}}(x,z)\int p_{t-\frac{k}{\mu}}(z,y_1)(K_{\mu-\beta}-K_\eta)(y_1)$$
$$\times\int p_{t-\frac{k}{\mu}}(z,y_2)\,(K_{\mu-\beta}-K_\eta)(y_2)\,dy_1dy_2.$$

Use the bounded density of m and the symmetry of p_t to integrate out x. Now use the scaling $p_t(x)=t^{-d/2}p_1(x/\sqrt{t})$, and the fact that $p_1(x)$ is bounded uniformly in x, to bound the above by a constant multiple of

$$\left(\frac{K}{\mu^2|m|}\right)\cdot\sum_{k=1}^{N}\left(t-\frac{k}{\mu}\right)^{-d}\int p_1\left(\frac{z}{\sqrt{t-k/\mu}},\frac{y_1}{\sqrt{t-k/\mu}}\right)dz$$
$$\times\int|K_{\mu-\beta}-K_\eta|(y_1)\,dy_1\int|K_{\mu-\beta}-K_\eta|(y_2)\,dy_2$$

Integrating out z gives factor of $(t-k/\mu)^{d/2}$. Note how nicely this breaks the time summation and the space integral into separate units, each of which can be handled separately. The two space integrals go to zero as $\mu\to\infty$ and $\eta\to 0$ by the $\mathcal{L}^2$ covergence of $K_{\mu-\beta}$ and K_η to G^λ. Thus all that one has to check is that

$$\lim_{\mu\to\infty}\left(\frac{K}{\mu^2|m|}\right)\cdot\sum_{k=1}^{N}\left(t-\frac{k}{\mu}\right)^{-d/2}<\infty,$$

keeping in mind, of course, that $K=O(\mu)$. This is left to the reader, to whom the advice of working with the limiting integral is offered.

This suffices for the first term of (4.8). We now consider the second term.

In fact, this is not much harder than the previous term. Write $Z_t^\mu(\phi)$ for $\int_0^t\phi(x)\,Z^\mu(ds,dx)$. Following the same procedure as before, write

$$Z_t^\mu(K_{\mu-\beta}) = Z_t^\mu(K_{\mu-\beta}-K_\eta) + Z_t^\mu(K_\eta),$$

and mimick the above argument, using now Theorem 2.2 for the weak convergence parts. Recall throughout that $\langle Z^\mu(\phi)\rangle_t=\int_0^t X_s^\mu(\phi^2)\,ds$ to do the second moment calculations. Details are left to the reader. (Note that since Theorem 2.2 only gives the weak convergence of $Z_t^\mu(\phi)$ for $\phi\in\mathcal{S}_d$, you will need an extra step, in which $K_{\mu-\beta}$ and K_η are approximated by such functions.) Overall, however, the calculation is not much harder than the above, since, after a few steps, the time sum and the space integral separate nicely, just as they did before.

To complete the proof one now only needs to put the two individual weak convergences we have just established together into the joint form of (4.8). If you have any doubts as to how this is done, see Chapter 7 of Walsh (1986). ∎

(b) Intersection local time. We start this subsection with the comparatively straightforward

PROOF OF THEOREM 2.5: What we need to prove here is that the weak limit, as $\mu \to \infty$, of

$$\gamma^\mu(\phi) = \frac{1}{\mu^2}\sum_\alpha\sum_{\beta\neq\alpha} L_T(\phi: X^\alpha, X^\beta),$$

where L_T is the intersection local time of Theorem 1.7, is also the $\mathcal{L}^2$ limit, as $\epsilon \to 0$, of

$$\gamma^\epsilon(\phi) := \int_0^T dt \int_0^t ds\, \langle \phi(x) f_\epsilon(x-y),\, X_s(dx)\, X_t(dy)\rangle, \tag{4.10}$$

for each $\phi \in \mathcal{S}_d$:

With the exception of one technical difficulty, the proof is very similar to that of Theorem 2.3 at the beginning of the previous subsection. Approximate γ^μ by γ^μ_ϵ, where

$$\gamma^\mu_\epsilon(\phi) := \frac{1}{\mu^2}\sum_\alpha\sum_{\beta\neq\alpha}\int_0^T dt\int_0^T ds\,\phi(X^\alpha(s))\, f_\epsilon(X^\alpha(s) - X^\beta(t)).$$

($\gamma^{\mu,\epsilon}$ would have been a nicer notation, but this has already been used for something else; cf. (2.10).)

Note that we can rewrite this as

$$\gamma^\mu_\epsilon(\phi) = \int_0^T dt\int_0^T ds \int_{\Re^d}\int_{\Re^d} \phi(x) f_\epsilon(x-y)\, X^\mu_s(dx) X^\mu_t(dy),$$

which, of course, is starting to look just like (4.10). What we now want to show is that

$$\lim_{\mu\to\infty}\lim_{\epsilon\to 0}\gamma^\mu_\epsilon(\phi) \stackrel{\mathcal{L}}{=} \lim_{\epsilon\to 0}\gamma_\epsilon(\phi).$$

This can be proven exactly as we showed in the proof above that

$$\lim_{\mu\to\infty}\lim_{\epsilon\to 0} L^{\mu,\epsilon}_t(x) \stackrel{\mathcal{L}}{=} \lim_{\epsilon\to 0} L^\epsilon_t(x),$$

once we know that, for fixed $\epsilon > 0$,

$$\gamma_\epsilon^\mu(\phi) \Rightarrow \gamma^\epsilon(\phi), \tag{4.11}$$

as $\mu \to \infty$. To show this, let $P^N(x,y)$ be a function of the form $\sum_{n=1}^N a_n\phi_n(x)\psi_n(y)$ such that $\|P^N(x,y)-\phi(x)\cdot f_\epsilon(x-y)\|_2 \to 0$ as $N \to \infty$, where the a_n are real and the ϕ_n and ψ_n are all bounded and of compact support. It is immediate that, for each finite $N > 0$,

$$\begin{aligned}\int_0^T dt \int_0^T ds \int_{\Re^d}\int_{\Re^d} & P^N(x,y)X_s^\mu(dx)X_t^\mu(dy) \\ &= \int_0^T dt \int_0^T ds \sum_{n=1}^N a_n X_s^\mu(\phi_n)X_t^\mu(\psi_n) \\ &\Rightarrow \int_0^T dt \int_0^T ds \sum_{n=1}^N a_n X_s(\phi_n)X_t(\psi_n) \\ &= \int_0^T dt \int_0^T ds \int_{\Re^d}\int_{\Re^d} P^N(x,y)X_s(dx)X_t(dy).\end{aligned}$$

This is almost (4.11). Going to the limit in N establishes (4.11), and completes (the outline of) the proof. ∎

We now start the longest and hardest proof of this paper: that of Theorem 2.6. As for the proof of Theorem 2.4, we require some preliminary lemmata.

4.3 LEMMA. *Let $\gamma^{\mu,\epsilon}(T,\phi)$ be the averaged, smeared, renormalised intersection local time of (2.10). Then, for all $\epsilon > 0$, $\lambda > 0$, $T > 0$ and $\phi \in \mathcal{S}_d$, $\gamma^{\mu,\epsilon}$ satisfies the evolution equation*

$$\begin{aligned}\gamma^{\mu,\epsilon}(T,\phi) &= \int_0^T\int_{\Re^d}\left\{\int_0^t \langle K_\epsilon(x-y)\phi(x),\, X_s^\mu(dx)\rangle\, ds\right\} Z^\mu(dt,dy) \\ &\quad + \lambda\int_0^T dt\int_0^t ds\,\langle K_\epsilon(x-y)\phi(x),\, X_s^\mu(dx)\, X_t^\mu(dy)\rangle \\ &\quad - \int_0^T \langle K_\epsilon(x-y)\phi(x),\, X_t^\mu(dx)\, X_T^\mu(dy)\rangle\, dt \\ &\quad + \frac{1}{\sqrt{\mu}}\int_0^T\int_0^t ds\int_{\Re^d}\int_{\Re^d}\nabla K_\epsilon(x-y)\phi(y)\, X_s^\mu(dy)W^\mu(dt,dx).\end{aligned} \tag{4.12}$$

PROOF: Fix $\epsilon > 0$, and indices α and β, so that either $\beta \prec \alpha$ or $\alpha \neq \beta$ if $|\alpha| = |\beta|$, and define the C^∞ function

$$F_\beta(t,x) = \int_0^t K_\epsilon(x - X_s^\beta)\phi(X_s^\beta)\, ds. \tag{4.13}$$

Note that because of our convention that $f(\Delta) \equiv 0$ for all functions f, the domain of the integral in (4.13) is really $[t \wedge |\beta|/\mu, t \wedge (|\beta|+1)/\mu]$. Now apply Itô's formula to the time inhomogeneous functional F_β to obtain that for all $T \sim \alpha$

(4.14)

$$\int_0^T K_\epsilon(X_T^\alpha - X_s^\beta)\phi(X_s^\beta)\, ds$$

$$= \int_0^{\frac{|\alpha|}{\mu}} K_\epsilon(X_{\frac{|\alpha|}{\mu}}^\alpha - X_s^\beta)\phi(X_s^\beta) ds + \int_{\frac{|\alpha|}{\mu}}^T \int_0^t \nabla K_\epsilon(X_t^\alpha - X_s^\beta)\phi(X_s^\beta) ds dY_t^\alpha$$

$$+ \int_{\frac{|\alpha|}{\mu}}^T K_\epsilon(X_t^\alpha - X_t^\beta)\phi(X_t^\beta)\, dt + \int_{\frac{|\alpha|}{\mu}}^T \int_0^t \Delta K_\epsilon(X_t^\alpha - X_s^\beta)\phi(X_s^\beta)\, ds dt.$$

So that (4.14) can be made to hold for all T, multiply the first term on the right hand side by $1_{[|\alpha|/\mu,\infty)}(T)$, and subtract, also from the right hand side,

$$1_{[(|\alpha|+1)/\mu,\infty)}(T) \lim_{t \uparrow \frac{|\alpha|+1}{\mu}} F_\beta\Big(\frac{|\alpha|+1}{\mu}, X_t^\alpha\Big).$$

Now apply (4.2) to rewrite the term with the Laplacian as the difference between two other terms, and rearrange everything so that you have only

$$\int_0^T dt \int_0^t f_\epsilon(X_s^\alpha - X_t^\beta)\,\phi(X_s^\beta)\, ds \; - \int_0^T K_\epsilon(X_t^\alpha - X_t^\beta)\phi(X_t^\beta)\, dt$$

on the left hand side of the equation. Sum both sides of the resulting equation over $\alpha \neq \beta$, note the definitions of $\gamma^{\mu,\epsilon}$, W^μ and Z^μ, keep in mind the convention about $f(\Delta)$, and the proof of the Lemma is complete. ∎

4.4 LEMMA. *Let $d \geq 3$, and let K_ϵ be as defined above. Then there exists a finite C such that, for all $1 \leq i \leq d$, and all $\epsilon > 0$,*

$$\left|\frac{\partial K_\epsilon(x)}{\partial x_i}\right| \leq \begin{cases} C\,|x_i| \cdot \|x\|^{-d} & \text{if } \|x\| \geq 2\epsilon \\ C\,\epsilon^{1-d} & \text{if } \|x\| < 2\epsilon. \end{cases} \tag{4.15}$$

PROOF: The key observation, which follows from the definition of the Green's function, is that

$$\begin{aligned}\frac{\partial G^\lambda(x)}{\partial x_i} &= \frac{\partial}{\partial x_i}\int_0^\infty \frac{e^{-\lambda t - \|x\|^2/2t}}{(2\pi)^{d/2} t^{d/2}}\, dt \\ &= -x_i \int_0^\infty \frac{e^{-\lambda t - \|x\|^2/2t}}{(2\pi)^{d/2} t^{(d+2)/2}}\, dt \\ &\leq C\,|x_i| \cdot \|x\|^{-d},\end{aligned}$$

with the last line following from the substitution $t \to t \cdot \|x\|^2$. We thus have

$$
\begin{aligned}
\left|\frac{\partial K_\epsilon(x)}{\partial x_i}\right| &= \left|\frac{\partial}{\partial x_i}\int_{\|y\|\le\epsilon} G^\lambda(x-y) f_\epsilon(y)\,dy\right| \\
&\le \int_{\|y\|\le 1}\left|\frac{\partial G^\lambda(x-\epsilon y)}{\partial x_i}\right| f(y)\,dy \qquad (4.16)\\
&\le C\int_{\|y\|\le 1}\frac{|x_i-\epsilon y_i|}{\|x-\epsilon y\|^d} f(y)\,dy
\end{aligned}
$$

Consider first the case $\|x\| \le 2\epsilon$. In this case, we can bound the last integral above by

$$
\epsilon^{-d} C \int_{\|y\|\le\epsilon}\frac{|x_i-y_i|}{\|x-y\|^d}\,dy \;\le\; \epsilon^{-d} C\int_{\|z\|\le 3\epsilon}\frac{|z_i|}{\|z\|^d}\,dz,
$$

which, after a conversion to polar coordinates, is easily seen to be bounded by ϵ^{1-d}.

For the case $\|x\| > 2\epsilon$, continue (4.16) with the bound

$$
C\frac{|x_i|}{\|x\|^d}\int_{\|y\|\le 1} f(y)\,dy \;=\; C\frac{|x_i|}{\|x\|^d}.
$$

This completes the proof. ∎

We now turn to the key Lemma required for the proof of Theorem 2.6.

4.5 Lemma. *Set $\epsilon = \mu^{-\beta}$, where $0 < \beta < 1/2$ if $d = 4$ and $0 < \beta < 1/3$ if $d = 5$. Assume that the initialising measure m has a bounded density with respect to Lebesgue measure. Then*

(4.17)

$$
E\Big\{\Big(\frac{1}{\sqrt{\mu}}\int_0^T\int_0^t ds\int_{\Re^d}\int_{\Re^d}\nabla K_\epsilon(x-y)\phi(y)\,X_s^\mu(dy)W^\mu(dt,dx)\Big)^2\Big\} \to 0
$$

as $\mu \to \infty$.

Proof: As in the proof of Lemma 4.2, we start with the known quadratic variation of the martingale measure W^μ (cf. (4.5)) to see that the expectation in (4.17) is equivalent to

$$
\begin{aligned}
&\frac{1}{\mu}E\Big\{\int_0^T dt\int_0^t ds_1\int_0^t ds_2\int_{\Re^{3d}}\nabla K_\epsilon(x-y_1)\nabla K_\epsilon(x-y_2)\\
&\qquad\times\phi(y_1)\phi(y_2)\,X_t^\mu(dx)X_{s_1}^\mu(dy_1)X_{s_2}^\mu(dy_2)\Big\}. \qquad (4.18)
\end{aligned}
$$

Interchanging expectation and the integral, we are now in a position to apply Theorem 3.4 to begin serious calculations. We make three simplifying claims. The first, rather bold one, is that it is "easy to see" that the $O(1/\mu)$ term referred to in Theorem 3.4 is of no importance to us. The boldness of this claim lies in the fact that the above expectation is really quite delicate, and will only be finite in the $\mu \to \infty$ limit if we chose β wisely. Nevertheless, we claim that if you follow the calculations below properly, then a little back of the envelope scribbling will convince you of our claim. A full proof takes a little longer, but is neither particularly difficult nor interesting.

The second claim is that although Theorem 3.1 indicates that we have a number of different diagrams to consider in calculating the inner expectation, it is enough to consider only one of them, since all the different diagrams have the same order of divergence in ϵ. The one that we shall look at in detail is the most complex.

The final claim, which is by far the easiest for the reader to verify, is that it makes no difference to the ultimate rate of convergence question if we replace the summations refered to in Theorem 3.4 by integrals. To check this, simply write out the following argument with summations instead of integrals, and at the very last stage go to the limiting integral. The only reason that we do not do this is that the following formulae, which are already long enough, become even more cumbersome.

The term (diagram) that we shall treat is the one that involves the following integral:

$$\frac{1}{\mu}\int_{\Re^d} m(dz)\int_0^T dt\int_0^t ds_1\int_0^t ds_2\int_{\Re^{5d}}\int_0^t dt_1\int_0^t dt_2 \\ \times \nabla K_\epsilon(x-y_1)\nabla K_\epsilon(x-y_2)\phi(y_1)\phi(y_2) \\ \times p_{t_1}(z,z_1)p_{t-t_1}(z_1,x)p_{t_2-t_1}(z_1,z_2)p_{s_2-t_2}(z_2,y_2)p_{s_1-t_2}(z_2,y_1),$$

where, to save on space, we have left out the 5 space differentials.

Use the fact that m has a bounded density, together with the symmetry of p, to integrate out z to remove the term $p_{t_1}(z,z_1)$. Then use Chapman-Kolmogorov to integrate out z_1, so that what remains is

$$\frac{1}{\mu}\int_0^T dt\int_0^t ds_1\int_0^t ds_2\int_{\Re^{4d}}\int_0^t dt_1\int_0^t dt_2\,\nabla K_\epsilon(x-y_1)\nabla K_\epsilon(x-y_2) \\ \times \phi(y_1)\phi(y_2)p_{t+t_2-2t_1}(z_2,x)p_{s_2-t_2}(z_2,y_2)p_{s_1-t_2}(z_2,y_1).$$

As before, use the scaling relationship $p_t(u) = t^{-d/2}p_1(u/\sqrt{t})$, and the fact that $p_1(u)$ is bounded uniformly in u to replace $p_{t+t_2-2t_1}(z_2,x)$ by $C(t+t_2-2t_1)^{-d/2}$. It is now possible to integrate out z_2 by Chapman-Kolmogorov. If we once again use the scaling relationship and the boundedness of p_1, we find that the time and space integrals above split to give

(4.19)
$$\frac{C}{\mu}\int_0^T dt\int_0^t ds_1\int_0^t ds_2\int_0^t dt_1\int_0^t dt_2\,\left((t+t_2-2t_1)(s_1+s_2-2t_2)\right)^{-d/2}$$
$$\times\int_{\Re^{3d}}\nabla K_\epsilon(x-y_1)\nabla K_\epsilon(x-y_2)\,\phi(y_1)\phi(y_2)\,dx dy_1 dy_2.$$

One can now check that the time integrals, which are independent of μ and ϵ, converge for all $d \le 5$. (Had we stayed with the time summations, as mentioned above, the entire argument would be identical up to this point, and one would merely have to show that the sums that approximate the time integrals in (4.19) converge to the right thing. The appropriate condition for this is the finiteness of the integrals; viz. $d \le 5$.)

We tackle the space integrals by starting with the region $\{y_1, y_2 \in \Re^d, \|x-y_1\| \vee \|x-y_2\| \le 2\epsilon\}$. Use (4.15) to bound each ∇K_ϵ by ϵ^{1-d}, and then note that since the x integral is now over a subset of a ball of radius 2ϵ, and therefore contributes a factor of ϵ^d, all that remains is

$$\frac{C\,\epsilon^{2-d}}{\mu}\int_{\Re^2}\phi(y_1)\phi(y_2)\,dy_1 dy_2.$$

The integral is finite, so that we have an overall bound of $O(\epsilon^{2-d})$

Over the two regions $\{\|x-y_1\| \le 2\epsilon,\ \|x-y_2\| \ge 2\epsilon\}$ and $\{\|x-y_1\| \ge 2\epsilon,\ \|x-y_2\| \le 2\epsilon\}$, use (4.15) and argue as follows, where we write $B_\epsilon(z)$ for the ϵ-ball centered on $z \in \Re^d$.

$$\frac{C\,\epsilon^{1-d}}{\mu}\int_{\Re^{2d}}\int_{\|x-y_1\|\le 2\epsilon,\,\|x-y_2\|\ge 2\epsilon}\frac{\sum_{i=1}^d |x(i)-y_2(i)|}{\|x-y_2\|^d}\,\phi(y_1)\phi(y_2)\,dx dy_1 dy_2$$
$$\le \frac{C\,\epsilon^{1-d}}{\mu}\int_{\Re^d}\phi(y_1)\,dy_1\int_{\Re^d}\phi(y_2)\,dy_2\int_{\|z\|\in B_{2\epsilon}(y_1-y_2)\cap B^c_{2\epsilon}(0)}\frac{\|z\|}{\|z\|^d}\,dz$$
$$\le \frac{C\,\epsilon^{1-d}\epsilon^d\epsilon^{1-d}}{\mu}$$

which is again of $O(\epsilon^{2-d})$.

The final region, in which $\{\|x-y_1\| \wedge \|x-y_2\| \ge 2\epsilon\}$, must be broken up into four subregions, depending on whether $\|x-y_1\|$ and $\|x-y_2\|$ are greater, or less than, 1. We leave it to the reader to check that the largest of these terms is again of $O(\epsilon^{2-d})$.

Since this now gives us that the expectation in (4.17) is $O(\mu^{-1}\epsilon^{2-d})$, setting $\epsilon = \mu^{-\beta}$, and noting the conditions on β, proves the Lemma. ∎

PROOF OF THEOREM 2.6: If we can show that all the terms on the right hand side of (4.12), which gives the evolution equation for the approximate SILT $\gamma^{\mu,\epsilon}(T,\phi)$ converge weakly (jointly) for each fixed T and ϕ to their obvious limits, appearing in the evolution equation (1.19) for the SILT, then we will be done.

We have already shown in Lemma 4.5 that the last, and most delicate term disappears in the limit under the conditions that we have assumed. It turns out that we no longer need the most stringent of these conditions: The dimensional restrictions that determine the existence of the SILT are all we now need.

Looking back at (4.12), there are three types of terms that we have to consider. Symbolically, these can be written as $\int_0^T \langle \int_0^t \langle \Psi, X_s^\mu \rangle ds, Z^\mu(dt) \rangle$, $\int_0^T \int_0^t \langle \Psi, X_s^\mu \times X_t^\mu \rangle \, dsdt$, and $\int_0^T \langle \Psi, X_s^\mu \times X_T^\mu \rangle \, ds$, where $\Psi : \Re^d \times \Re^d \to \Re$ actually depends on ϵ and, therefore, μ.

If the test function Ψ were μ-independent, then each of these terms would be quite straightforward to handle. We could, as in the proof of Theorem 2.5, approximate Ψ in $\mathcal{L}^2$ by a function of the form $\sum_{n=1}^N a_n \phi_n(x) \psi_n(y)$, for which passage to the limit is simple. The proof is then completed as for Theorems 2.4 and 2.5.

As we noted above, one of the reasons that we do not want to go through all the technical details of the proof is that we hope to have a "meta-theorem" for these kinds of problems written up soon, and this paper is already long enough as is, without adding proofs that are soon to be outdated.

5. REMAINING PROOF

All that remains is to tie up one loose end.

Proof of (1.17): Since it is late in the day, and the Bass/Burdzy submission deadline expired a little over a month ago, I shall cheat a little and only prove the approximate version of (1.17), viz.

$$\begin{aligned} SILT_\epsilon(B) &= \int_B dsdt \int_{\Re^d \times \Re^d} f_\epsilon(x-y) X_s(dx) X_t(dy) \\ &= \int_{t_0}^{t_1} \int_{t_2}^{t_3} \langle K_\epsilon(x-y), X_{s_2}(dx) \rangle \, Z(ds_1, dy) ds_2 \\ &\quad + \lambda \int_{t_0}^{t_1} \int_{t_2}^{t_3} \langle K_\epsilon(x-y), X_{s_1}(dx) X_{s_2}(dy) \rangle \, ds_1 ds_2 \\ &\quad - \int_{t_2}^{t_3} ds_2 \int_{\Re^d} \langle K_\epsilon(x-y), X_{s_2}(dy) \rangle \, (X_{t_1} - X_{t_0})(dx). \end{aligned} \tag{5.1}$$

Here, with a slight extension (to general α) of the notation of the previous section, $K_\epsilon := G_\alpha^\lambda * f_\epsilon$, and so satisfies the equation

$$(\Delta_\alpha + \lambda) K_\epsilon = f_\epsilon. \tag{5.2}$$

To see how to send $\epsilon \to 0$ for the $\mathcal{L}^2$ convergence of each term in (5.1) and a full proof of (1.17) you will have to follow the moment argument of Adler and Lewin (1991).

To derive (5.1), start with the first line and use (5.2) to obtain

$$\begin{aligned} SILT_\epsilon(B) &= \lambda \int_{t_0}^{t_1} ds_1 \int_{t_1}^{t_2} ds_2 K_\epsilon(x-y)\, X_{s_1}(dx) X_{s_2}(dy) \\ &\quad - \int_{t_0}^{t_1} ds_1 \int_{t_1}^{t_2} ds_2 \Delta_\alpha K_\epsilon(x-y)\, X_{s_1}(dx) X_{s_2}(dy) \end{aligned} \tag{5.3}$$

The first term, modulo a minor change of notation, already appears in the last line of (5.1). To handle the second, recall the fundamental SPDE (1.2a) to write, for each fixed $y \in \Re^d$,

$$\begin{aligned} &\int_{t_0}^{t_1} ds_1 \int_{\Re^d} \Delta_\alpha K_\epsilon(x-y)\, X_{s_1}(dx) \\ &\quad = \int_{\Re^d} K_\epsilon(x-y)\,(X_{t_1} - X_{t_0})(dx) \;-\; \int_{t_0}^{t_1} \int_{\Re^d} K_\epsilon(x-y)\, Z(ds_1, dy). \end{aligned}$$

Integrate both sides of the above against $X_{s_2}(dy)$, integrate s_2 over $[t_2, t_3]$, and substitute back into (5.3) to obtain (5.1), thus completing the proof. ∎

ACKNOWLEDGEMENT: I am grateful to Raisa Epstein Feldman, Srikanth Iyer, Steve Krone and the ubiquitous referee for a number of useful comments on, and for finding a number of minor errors in, previous versions of this paper. My debt of gratitude to Ed Perkins for his notes Perkins (1992) is major.

6. REFERENCES

[1] Adler, R.J. and Epstein, R. (1987), A central limit theorem for Markov paths and some properties of Gaussian random fields, *Stochastic Process Appl.* **25**, 157-202.

[2] Adler, R.J. Feldman R. and Lewin, M. (1991), Intersection local times for infinite systems of planar Brownian motions and the Brownian density process, *Ann. Probab.* **19**, 192-220.

[3] Adler, R.J. and Lewin, M. (1991), An evolution equation for the intersection local times for super processes, *Stochastic Analysis*, eds. M.T. Barlow and N.H. Bingham, Cambridge University Press, 1-22.

[4] Adler, R.J. and Lewin, M. (1992), Local time and Tanaka formulae for super Brownian motion and super stable processes, *Stochastic Process Appl.* **41**, 45-68.

[5] Adler, R.J. and Rosen, J.S. (1992) Intersection local times of all orders for Brownian and stable density processes – construction, renormalisation, and limit laws, *Ann. Probab.*, **20**, to appear.

[6] Bass, R.F. and Khoshnevisan, D. (1992) Intersection local times and Tanaka formulas, preprint.

[7] Blumenthal, R.M. and Getoor, R.K. (1968) *Markov Processes and Potential Theory*, New York, Academic.

[8] Dawson, D.A. Iscoe, I. and Perkins, E.A. (1989), Super Brownian motion: Path properties and hitting probabilities, *Prob. Theory and Related Topics*, **83**, 135–205.

[9] Dawson, D.A. and Hochberg, K.J. (1979), The carrying dimension of a stochastic measure diffusion, *Ann. Probability* **7**, 693-703.

[10] Dvoretzky, A. Erdös, P. and Kakutani, S. (1950), Double points of paths of Brownian motion in n-space, *Acta Sci. Math. (Szeged)* **12** 75-81.

[11] Dynkin, E.B. (1988a), Self-intersection gauge for random walks and for Brownian motion. *Ann. Probability,* **16** 1-57.

[12] Dynkin, E.B. (1988b) Regularised self–intersection local times of planar Brownian motion. *Ann. Probability* **16**, 58-74.

[13] Dynkin, E.B. (1988c), Representation for functionals of superprocesses by multiple stochastic integrals, with applications to self intersection local times, *Astérisque*, **157–158**, 147–171.

[14] Ethier, S. and Kurtz, T. (1986), *Markov Processes*, Wiley, New York.

[15] Fleischmann, K. (1988), Critical behavior of some measure valued processes, *Math. Nachr.* **135** 131-147.

[16] Geman, D. and Horowitz, J. (1980), Occupation densities *Ann. Probability* **8** 1-67.

[17] Glimm, J. and Jaffe, A. (1987) *Quantum Physics*, Springer-Verlag, New York.

[18] Iscoe, I. (1986a), A weighted occupation time for a class of measure valued branching processes, *Z. Wahrsch. verw. Gebeite* **71** 85–116.

[19] Iscoe, I. (1986b), Ergodic theory and a local occupation time for measure valued critical branching Brownian motion, *Stochastics* **18** 197–243.

[20] Krone, S. (1991), Local times for superdiffusions, (Abstract), *Stoch. Proc. Appl.* **35** 199-200.

[21] Le Gall, J-F. (1990), Some properties of planar Brownian motion, *XXième Ecole d'été de Probabilités de St-Flour.*

[22] Perkins, E. (1988) A space-time property of a class of measure valued branching diffusions, *Trans. Am. Math. Soc.* **305** 743-795.

[23] Perkins, E. (1990) *Lecture Notes on Superprocesses*, Handwritten (but extremely informative) manuscript.

[24] Revuz, D. and Yor, M. (1991) *Continuous Martingales and Brownian Motion*. New York: Springer.

[25] Rosen, J. (1986), Tanaka's formula and renormalisation for intersec-

tions of planar Brownian motions, *Ann. Probability*, **14** 1245-1251.

[26] Rosen, J. (1987), Joint continuity of intersection local times of Markov processes, *Ann. Probability*, **15** 659-675.

[27] Rosen, J. (1986) A renormalised local time for multiple intersections of planar Brownian motion. In: Séminaire de Probabilités XX (eds. Azèma, J., Yor, M.), 515–531. New York: Springer.

[28] Rosen, J. (1990a) Random walks and intersection local time. *Ann. Probabability* **18**, 959–977.

[29] Rosen, J. (1990b), Renormalisation and limit theorems for self-intersections of super-processes. Preprint.

[30] Rosen, J. and Yor, M. (1991) Tanaka formulae and renormalisation for triple intersections of Brownian motion in the plane. *Ann. Probability* **19**, 142–159.

[31] Sugitani, S. (1988) Some properties for the measure-valued branching diffusion process, *J. Math. Soc. Japan* **41** 437–462.

[32] Taylor, S.J. (1986) The measure theory of random fractals, *Math. Proc. Cambridge Phil. Soc.* **100**, 383-406.

[33] Varadhan, S. (1969), Appendix to: Euclidean quantum field theory, by K. Symanzik. In *Local Quantum Field Theory* (R. Jost ed.) Academic, New York.

[34] Walsh, J.B. (1986), An introduction to stochastic partial differential equations, *Springer Lecture Notes in Math.*, **1180**, 265-439.

[35] Watanabe, S. (1968), A limit theorem for branching processes and continuous state branching processes, *J. Math. Kyoto Univ.* **8** 141–167.

[36] Yor, M. (1985) Compléments aux formules de Tanaka–Rosen. In: Séminaire de Probabilités XIX (eds. Azèma, J., Yor, M.), 332–349. New York: Springer.

[37] Yoshida, K.Y. (1980), *Functional Analysis*, Springer-Verlag, Berlin.

Faculty of Industrial Engineering and Management
Technion – Israel Institute of Technology
Haifa ISRAEL 32000
IERHE01@TECHNION.TECHNION.AC.IL

Strong Approximations to Brownian Local Time

RICHARD F. BASS [1] and DAVAR KHOSHNEVISAN

1. Introduction

A strong approximation (or strong invariance principle) of random walks to Brownian motion is the construction of random walks and Brownian motions on a single probability space so that the normalized random walks converge to the Brownian motions a.s. at some rate. Given such a strong approximation, it is natural to ask if the "local times" of the random walks also converge a.s. to the local times of the Brownian motions. The first such strong invariance principle for local times that also gave a rate was given in [R] for the case of simple random walk. Since then there have been many papers on this subject with the aim of weakening the assumptions on the walk and improving the rate. See [Bo2] and [CH] and the references therein. For recent results along these lines, see [BK2].

Our aim in this paper is slightly different. We are not concerned with obtaining optimal rates here, but rather giving the most general conditions under which a strong approximation for the local times holds. We assume that our random walks are mean 0 and variance 1, but no other moment conditions are assumed. Our main results (Theorems 3.2 and 4.5) say that one has a strong approximation for the local times whenever one has a strong invariance principle for the random walks, provided only that some very mild conditions hold. This is

[1] Research partially supported by NSF grant

quite surprising because local time is a highly discontinuous functional on $C[0,1]$. In particular this shows that the local times of the random walks converge a.s. to those of the Brownian motions in the case of the Hungarian embedding. Our results also give a strong invariance principle for random walks having only $2+\varepsilon$ moments.

The conditions we impose are (a) that either the random walk be lattice or strongly nonlattice and (b) that the rate of convergence of the random walks to the Brownian motions not be too slow.

There are a number of interesting by-products of our results. We obtain weak uniform invariance principles for the local times of finite variance random walks in the lattice and strongly nonlattice case. The lattice case was already known ([Bo1]), while the strongly nonlattice case improves upon some results of Borodin (see [Bo2]). Our methods also give a weak uniform invariance principle for intersection local times with only second moments (this has also recently been proved by Rosen) and a strong uniform invariance principle for intersection local times, which is new. Finally, we show that the general theorems in [BK1] for weak uniform invariance principles for local times on curves hold under the assumption of finite variance when the dimension is 3 or less.

We use the letter c with or without subscripts to denote constants whose value is unimportant and may change from line to line.

2. A local central limit theorem

We begin with some estimates on the transition densities for random walks in $\mathbb{R}^d$. Let X_i be a sequence of i.i.d. random variables with mean 0 and nondegenerate covariance matrix Q. We assume the X_i are lattice-valued, and without loss of generality we assume the lattice to be $\mathbb{Z}^d$. Let $S_n = \sum_{i=1}^n X_i$ and let $P_n(x,y) = \mathbb{P}^x(S_n = y)$. For now let us assume the S_n are strongly aperiodic (see [Sp, D5.1]).

Proposition 2.1. *Let* $\gamma \in (0,1]$. *Then*

$$|P_n(0,x) - P_n(0,y)| \le cn^{-d/2}(|x-y|/\sqrt{n})^\gamma. \tag{2.1}$$

Proof. Following Spitzer ([Sp, T7.9]),

$$\begin{aligned} &P_n(0,x) - P_n(0,y) \\ &\quad = cn^{-d/2} \int_{\alpha \in \sqrt{n}C} \varphi^n(\alpha/\sqrt{n})(e^{-ix\cdot\alpha/\sqrt{n}} - e^{-iy\cdot\alpha/\sqrt{n}})\, d\alpha, \end{aligned}$$

where φ is the characteristic function of X_i and C is the cube of side length 2π centered at the origin.

Let $\gamma \in (0,1]$ and note that $|e^{ia} - e^{ib}| \le c|a-b|^\gamma$. Let $\Delta = (|x-y|/\sqrt{n})^\gamma$. For $|\alpha| \le 1$, we have $|\varphi(\alpha/\sqrt{n})| \le 1$, and so

$$\int_{|\alpha|\le 1} \le c\Delta \int_{|\alpha|\le 1} |\alpha|^\gamma \, d\alpha \le c\Delta.$$

For r small and $|\alpha| \in [1, r\sqrt{n}]$, we have $|\varphi^n(\alpha/\sqrt{n})| \le \exp(-\alpha^t Q\alpha/4)$ (see [Sp, p. 77]), where α^t denotes the transpose of α. So

$$\int_{1\le|\alpha|\le r\sqrt{n}} \le c\Delta \int_{1\le|\alpha|\le r\sqrt{n}} |\alpha|^\gamma \exp(-\alpha^t Q\alpha/4)\, d\alpha \le c\Delta.$$

Since the X_i are strongly aperiodic, $|\varphi^n(\alpha/\sqrt{n})| \le (1-\delta)^n$ for some $\delta > 0$ if $|\alpha| \ge r\sqrt{n}$ (cf. [Sp, p. 77]). Hence for any $K > 0$

there exists c such that

$$\int_{r\sqrt{n}\leq|\alpha|} \leq c\Delta(1-\delta)^n \int_{r\sqrt{n}\leq|\alpha|\leq\pi\sqrt{n}} |\alpha|^\gamma \, d\alpha \leq c\Delta n^{-K}.$$

Summing the three terms gives (2.1). □

A similar proof, following [Sp, P7.10], shows

$$\left|\frac{|x|^2}{n}P_n(0,x) - \frac{|y|^2}{n}P_n(0,y)\right| \leq cn^{-d/2}(|x-y|/\sqrt{n})^\gamma. \tag{2.2}$$

Let $G(x,y) = \sum_{n=0}^\infty P_n(x,y)$.

Proposition 2.2. *Suppose $d = 3$. Let $\gamma \in (0,1)$. Then*

$$|G(0,x) - G(0,y)| \leq c\frac{|x-y|}{(|x|\wedge|y|)^2} + c\frac{|x-y|^\gamma}{(|x|\wedge|y|)^{1+\gamma}}.$$

Proof. Suppose $|x| \leq |y|$. Note

$$\begin{aligned} P_n(0,x) - P_n(0,y) = \frac{n}{|x|^2}\left(\frac{|x|^2}{n}P_n(0,x) - \frac{|y|^2}{n}P_n(0,y)\right) \\ + \left(\frac{|y|^2}{|x|^2} - 1\right)P_n(0,y). \end{aligned}$$

Using (2.2) and the fact that $P_n(0,y) \le n^{-1/2}|y|^2$ ([Sp, P7.10]),

$$\begin{aligned}\sum_{n=1}^{[|x|^2]} |P_n(0,x) - P_n(0,y)| &\le c \sum_{n=1}^{[|x|^2]} \frac{n}{|x|^2} \frac{1}{n^{3/2}} \frac{|x-y|^\gamma}{n^{\gamma/2}} \\ &\quad + c\frac{|y|^2 - |x|^2}{|x|^2} \sum_{n=1}^{[|x|^2]} \frac{1}{n^{1/2}|y|^2} \\ &\le c\frac{|x-y|^\gamma}{|x|^2}|x|^{1-\gamma} \\ &\quad + c\frac{(|y|-|x|)(|y|+|x|)}{|x|^2|y|^2}|x| \\ &\le c\frac{|x-y|^\gamma}{|x|^{1+\gamma}} + c\frac{|x-y|}{|x|^2}.\end{aligned}$$

Using Proposition 2.1,

$$\sum_{n=[|x|^2]}^{\infty} |P_n(0,x) - P_n(0,y)| \le c\frac{|x-y|^\gamma}{|x|^{1+\gamma}}.$$

Adding these two inequalities gives our result. □

Using the method of [Sp, p. 310], the assumption of strong aperiodicity may be dispensed with. Note Proposition 2.2 and [Sp, P26.1] together give us the conclusion of Corollary 3.3 of [BK1] when $d = 3$ under the assumption of finite variance only. By the proofs of [BK1], then, we get

Theorem 2.3. *Suppose $d \le 3$, the X_i's have finite variance, and Hypothesis 6.1 of* [BK1] *holds for some $\beta \in (0,1)$. Then the conclusion of Proposition 6.3 of* [BK1] *holds.*

Remarks. (1) The importance of Theorem 2.3 to us here is that Theorem 7.4 of [BK1] holds with only second moments, i.e., the local times of the random walk converge weakly to the local times of Brownian motion, uniformly over all levels x. This had previously been proved in [Bo1].

(2) Theorem 2.3 also tells us that Theorem 7.8 of [BK1] holds with only second moments, i.e., the intersection local times of two independent random walks on $\mathbb{Z}^2$ or $\mathbb{Z}^3$ converges weakly to the intersection local times of two independent Brownian motions, uniformly over all levels x. While we were writing up the present results, we learned that Jay Rosen had independently proved a more general result about weak uniform invariance principles for intersection local times. This did not surprise us – it was through discussions with him that we came to believe Theorem 7.8 of [BK1] should hold with only second moments.

3. Local times – lattice case

We assume in this section that we have $\mathbb{Z}$–valued random walks with mean 0, variance 1. We also assume that our walks are strongly aperiodic – this assumption can be dispensed with by the use of [Sp, P5.1].

Define $\eta(x,n) = \sum_{j=1}^{n} 1_{\{x\}}(S_j)$, the number of times before n that the random walk hits x. Let $\mathcal{F}_j$ be the σ–field generated by $S_1, \ldots, S_j$. Let

$$L^n(x,t) = \frac{1}{\sqrt{n}}\eta(\sqrt{n}x, nt)$$

for $x \in \mathbb{Z}/\sqrt{n}$, $t \in \mathbb{Z}/n$, and let $L^n(x,t)$ be defined by linear interpolation for all other x and t.

Proposition 3.1. *Let $\delta \in (0, 1/2)$. Then there exist*

c_1, c_2 such that

$$\mathbb{P}\Big(\sup_{x,y\in\mathbb{R},t\le 1} |L^n(x,t) - L^n(y,t)|/(|x-y|^{1/2-\delta} \wedge 1) \ge \lambda\Big) \le c_1 \exp(-c_2\lambda) + n^{-7}$$

and for large n (depending on ω)

$$\sup_{x,y\in\mathbb{R},t\le 1} |L^n(x,t) - L^n(y,t)|/(|x-y|^{1/2-\delta} \wedge 1) \le c\log n, \quad a.s.$$

Proof. Fix $x, y \in \mathbb{Z}$, $x \ne y$, and define $A_k = (\eta(x,k) - \eta(y,k))/\sqrt{n}$.

By Proposition 2.1 and the translation invariance of the P_j,

$$(3.1) \qquad |\mathbb{E}^z A_k| \le \frac{1}{\sqrt{n}} \sum_{j=1}^{k} |P_j(z,x) - P_j(z,y)| \le \frac{1}{\sqrt{n}} \sum_{j=1}^{k} \frac{c}{\sqrt{j}} \left(\frac{|x-y|}{\sqrt{j}} \right)^{\gamma} = c\frac{|x-y|^{\gamma}}{n^{\gamma/2}}.$$

Similarly, using [Sp, P7.9],

$$\mathbb{E}^z \eta(x,k) \le c\sqrt{n}.$$

Let $\Delta A_j = A_{j+1} - A_j$. Since $|x-y| \ge 1$,

$$\mathbb{E}^z \sum_{j=0}^{n-1} (\Delta A_j)^2 \le \frac{1}{n} \mathbb{E}^z \sum_{j=0}^{n-1} 1_{\{x,y\}}(S_j)$$

$$= \frac{1}{n}(\mathbb{E}^z \eta(x,n) + \mathbb{E}^z \eta(y,n)) \tag{3.2}$$

$$\le \frac{c}{\sqrt{n}} \le c\frac{|x-y|}{\sqrt{n}}.$$

Now

$$\mathbb{E}^z \sum_{j=0}^{n-1}(A_n - A_{j+1})\Delta A_j = \mathbb{E}^z \left[\sum_{j=0}^{n-1} \mathbb{E}^z (A_n - A_{j+1} | \mathcal{F}_{j+1})\Delta A_j \right]$$

$$= \mathbb{E}^z \sum_{j=0}^{n-1} \mathbb{E}^{S_{j+1}}(A_{n-j-1})\Delta A_j \tag{3.3}$$

$$\le c\frac{|x-y|^\gamma}{n^{\gamma/2}} \mathbb{E}^z \sum_{j=0}^{n-1} |\Delta A_j|$$

$$\le c\frac{|x-y|^\gamma}{n^{\gamma/2}} \frac{1}{\sqrt{n}} \mathbb{E}^z [\eta(x,n) + \eta(y,n)]$$

$$\le c\left(\frac{|x-y|}{\sqrt{n}}\right)^\gamma.$$

Since

$$A_n^2 = 2\sum_{j=0}^{n-1}(A_n - A_{j+1})\Delta A_j + \sum_{j=0}^{n-1}(\Delta A_j)^2,$$

(cf. [BK1, Eq. (5.3)]), combining (3.2) and (3.3) gives

$$\sup_z \mathbb{E}^z A_n^2 \le c\Delta,$$

where

$$\Delta = \left(\frac{|x-y|}{\sqrt{n}}\right)^\gamma + \left(\frac{|x-y|}{\sqrt{n}}\right).$$

Let

$$U_j = \mathbb{E}^z[A_n - A_j | \mathcal{F}_j] = \mathbb{E}^{S_j} A_{n-j}.$$

By (3.1), $|U_j| \leq c\Delta$. Since $M_j = U_j + A_j$ is a martingale, Doob's inequality tells us that

$$\begin{aligned}
\mathbb{E}^z(\sup_{j\leq n} |A_j|)^2 &\leq 2\mathbb{E}^z(\sup_{j\leq n} |U_j|)^2 + 2\mathbb{E}^z(\sup_{j\leq n} |M_j|)^2 \\
&\leq c\Delta^2 + 8\mathbb{E}^z M_n^2 \\
&\leq c\Delta^2 + 16\mathbb{E}^z U_n^2 + 16\mathbb{E}^z A_n^2 \\
&\leq c(\Delta^2 + \Delta).
\end{aligned}$$

Hence for any $\mathcal{F}_j$–stopping time N bounded by n,

$$\mathbb{E}^z[(A_n - A_N)^2 | \mathcal{F}_N] \leq \mathbb{E}^{S_N}[\sup_{j\leq n} |A_j|^2] \leq c(\Delta^2 + \Delta).$$

Using Cauchy–Schwarz, the fact that $\sup_{j\leq n-1} |\Delta A_j| \leq 1/\sqrt{n}$, and [DM, p. 193],

$$\mathbb{P}^z(\sup_{k\leq n} |A_k| \geq \lambda) \leq c_1 \exp(-c_2\lambda/(\Delta + \Delta^2)^{1/2}) \tag{3.4}$$

(cf. [BK1, Proposition 5.2]).

A standard metric entropy argument using (3.4) shows that for each $\delta > 0$,

$$\begin{aligned}
\mathbb{P}^z(\sup_{|x|,|y|\leq n^4, t\leq 1} &|L^n(x,t) - L^n(y,t)|/(|x-y|^{1/2-\delta} \wedge 1) > \lambda) \\
&\leq c_1 \exp(-c_2\lambda).
\end{aligned}$$

Using the fact that

$$\mathbb{P}^0(\sup_{j\leq n} |S_j| \geq n^4) \leq n^{-7} \tag{3.5}$$

by Chebyshev's inequality, we get the first assertion. A standard Borel–Cantelli argument (with $\lambda_n = c\log n$) shows that there exists c such that for large n (depending on ω)

(3.6)

$$\sup_{x,y\in\mathbb{R},t\leq 1} |L^n(x,t) - L^n(y,t)|/(|x-y|^{1/2-\delta} \wedge 1) \leq c\log n, \quad \text{a.s.}$$

This completes the proof. □

We are now ready for the main result of this section. Suppose for each n we have i.i.d. random variables $X_1^n, \ldots, X_n^n$, and we set $S_k^n = \sum_{j=1}^k X_j^n$. Define η^n in terms of the S_j^n analogously to the definition of η above and let us now use $L^n(x,t)$ to denote $\eta^n(\sqrt{n}x, nt)/\sqrt{n}$ for $x \in \mathbb{Z}/\sqrt{n}$, $t \in \mathbb{Z}/n$, with linear interpolation for all other x and t. For each n let B_t^n be a Brownian motion with local times $\ell^n(x,t)$.

Theorem 3.2. *Let $\sigma > 0$. Suppose a_n is a sequence such that for some $\delta > 0$, $a_n \leq (\log n)^{-6}$. Suppose*

$$\sup_{t\leq 1} |S_{[nt]}^n/\sqrt{n} - B_t^n| = O(a_n), \quad a.s.$$

Then

$$\sup_{x\in\mathbb{R},t\leq 1} |L^n(x,t) - \ell^n(x,t)|$$
$$= O(n^{-1/(5+\sigma)} \wedge a_n^{1/(5+\sigma)} \log n), \quad a.s.$$

Remarks. (1) The statement of our theorem includes the case where the X_j^n do not depend on n, B_t is a single Brownian motion, and $B_t^n = B_{nt}/\sqrt{n}$, the situation that comes up in Skorokhod embedding.

(2) Much better rates are possible when the X_i satisfy higher moment conditions. See [BK2].

(3) If the X_j have $2+\varepsilon$ moments, then any reasonable Skorokhod embedding will have a_n satisfying the above hypotheses.

(4) The Hungarian embedding satisfies the above with $a_n = \log n/\sqrt{n}$.

Proof. First note that (3.6) holds for the new $L^n(x,t)$ as well as the old ones, since the estimates (3.4) and (3.5) are valid with constants independent of n. Similarly we get the estimates

$$\sup_{t\le 1} |\ell^n(x,t) - \ell^n(y,t)| \le c(|x-y|^{1/2-\delta} \wedge 1) \log n, \quad \text{a.s.} \tag{3.7}$$

and

$$\sup_{x\in\mathbb{R}, t\le 1} L^n(x,t) \le c \log n, \quad \text{a.s.} \tag{3.8}$$

for large n (depending on ω).

We now fix an ω not in any of the exceptional sets and proceed to get a bound on $|L^n(x,t) - \ell^n(x,t)|$ independent of x and t. For simplicity of notation we will do the case $t = 1$. Fix x_0. Define $f_\varepsilon(x_0) = 1/\varepsilon$, define $f_\varepsilon(y) = 0$ if $|y - x_0| \ge \varepsilon$, and define it by linear interpolation for $y \in (x_0 - \varepsilon, x_0)$ and $(x_0, x_0 + \varepsilon)$. Take δ small.

Since $\int f_\varepsilon(y)\,dy = 1$,

$$\begin{aligned} |\int_0^1 f_\varepsilon(B_s^n)\,ds - \ell^n(x_0,1)| &= |\int f_\varepsilon(y)\ell^n(y,1)\,dy - \ell^n(x_0,1)| \\ &\le \sup_{|y-x_0|\le\varepsilon} |\ell^n(y,1) - \ell^n(x_0,1)| \\ &\le c\varepsilon^{1/2-\delta} \log n. \end{aligned} \tag{3.9}$$

Next,

$$
\begin{aligned}
&\left|\int_0^1 f_\varepsilon(B^n_s)\,ds - \frac{1}{n}\sum_{j=1}^n f_\varepsilon(S_j/\sqrt{n})\right| \\
&\le \left|\int_0^1 f_\varepsilon(B^n_s)\,ds - \int_0^1 f_\varepsilon(S_{[ns]}/\sqrt{n})\,ds\right| + 2/\varepsilon n \qquad (3.10)\\
&\le \|f'_\varepsilon\|_\infty a_n + 2/\varepsilon n = a_n/\varepsilon^2 + 2/\varepsilon n.
\end{aligned}
$$

Now

$$
\begin{aligned}
\frac{1}{n}\sum_{j=1}^n f_\varepsilon(S_j/\sqrt{n}) &= \frac{1}{n}\sum_{y\in\mathbb{Z}/\sqrt{n}} f_\varepsilon(y)\eta^n(\sqrt{n}y, n)\\
&= \frac{1}{\sqrt{n}}\sum_{y\in\mathbb{Z}/\sqrt{n}} f_\varepsilon(y)[L^n(y,1) - L^n(x_0,1)]\\
&\quad + L^n(x_0,1)\left[\frac{1}{\sqrt{n}}\sum_{y\in\mathbb{Z}/\sqrt{n}} f_\varepsilon(y) - 1\right] + L^n(x_0,1).
\end{aligned}
$$

It is easy to see that $|n^{-1/2}\sum_{y\in\mathbb{Z}/\sqrt{n}} f_\varepsilon(y) - 1| \le 1/n\varepsilon$. So using (3.6) and (3.8),

$$
\begin{aligned}
&\left|\frac{1}{n}\sum_{j=1}^n f_\varepsilon(S_j/\sqrt{n}) - L^n(x_0,1)\right| \\
&\le c\varepsilon^{1/2-\delta}\log n + c\log n/\varepsilon n. \qquad (3.11)
\end{aligned}
$$

Adding (3.9), (3.10) and (3.11) we get that

$$
|L^n(x_0,1) - \ell^n(x_0,1)| \le c\varepsilon^{1/2-\delta}\log n + a_n/\varepsilon^2 + c\log n/\varepsilon n.
$$

Now choose $\varepsilon = \varepsilon_n = a_n^{2/5} \vee n^{-1/2}$. $\square$

Corollary 3.3. *Let $\sigma > 0$. If $a_n \to 0$ and*

$$\sup_{t\le 1} |S^n_{[nt]}/\sqrt{n} - B^n_t| = O(a_n) \quad \textit{in probability,}$$

then

$$\sup_{x\in\mathbb{R},t\le 1} |L^n(x,t) - \ell^n(x,t)|$$
$$= O(a_N^{1/(5+\sigma)} \wedge n^{-1/(5+\sigma)}) \quad \textit{in probability.}$$

Proof. Let δ be chosen small. Let $\zeta > 0$. Provided the constants c are chosen large enough, we have by the remarks preceding Theorem 3.2 and the proof of Theorem 3.2 that for all n sufficiently large,

$$\mathbb{P}(\sup_{x,y\in\mathbb{R},t\le 1} |L^n(x,t) - L^n(y,t)|/(|x-y|^{1/2-\delta} \wedge 1) > c) < \zeta/3,$$

$$\mathbb{P}(\sup_{x,y\in\mathbb{R},t\le 1} |\ell^n(x,t) - \ell^n(y,t)|/(|x-y|^{1/2-\delta} \wedge 1) > c) < \zeta/3$$

and

$$\mathbb{P}(\sup_{t\le 1,x\in\mathbb{R}} |L^n(x,t)| > c) < \zeta/3.$$

So, following the proof of Theorem 3.2, for each n sufficiently large there exists a set N_n of probability less than ζ such that if $\omega \notin N_n$, then

$$|L^n(x_0,1) - \ell^n(x_0,1)| \le c\varepsilon^{1/2-\delta} + a_n/\varepsilon^2 + c/\varepsilon n$$

(the $\log n$'s are no longer present). Now choose ε_n just as before. □

Remark. If one has a lattice valued random walk in two or three dimensions, strong approximations to 2 or 3 dimensional Brownian motion exist (see [E]). A proof very similar to the one above shows that if two independent random walks converge uniformly to two independent Brownian motions a.s. at a sufficiently fast rate a_n, then the intersection local times of the two random walks will converge a.s. to the intersection local times of the two Brownian motions, uniformly over all levels x, at a rate of a_n to a suitable power.

4. Local times – nonlattice case

In this section we assume the X_i are i.i.d. $\mathbb{R}$–valued random variables with mean 0, variance 1 and are nonlattice, i.e., $|\varphi(\theta)| < 1$ for all $\theta \neq 0$. We also make the assumption throughout this section that the X_i are strongly nonlattice, that is,

$$\limsup_{|\theta|\to\infty} |\varphi(\theta)| < 1.$$

Since $|\varphi(\theta)| < 1$ for all $\theta \neq 0$, we get in this case that for any $r > 0$

$$\rho_r = \sup_{|\theta|\geq r} |\varphi(\theta)| < 1.$$

The assumption that the X_i are strongly nonlattice is reasonably weak, being satisfied, for example, if the distribution of X_i has a nonzero absolutely continuous part. (This follows using the Riemann–Lebesgue lemma.) In particular, Borodin's condition [Bo2], that $\varphi \in L^2$, implies that S_2 has a bounded density, hence that $\varphi(\theta)^2 \to 0$ as $|\theta| \to \infty$. See also Section 4 of [BK2].

We proceed to get two estimates of local limit type. Stone [St] has similar results without the strongly nonlattice assumption, but they are not sufficiently strong enough for our purposes here.

Let I_h denote $[-h, h]$.

Proposition 4.1. *Let $K \in \mathbb{Z}^+$. Then for all x, z*

$$\mathbb{P}^z(S_n \in x + I_h) \leq c\frac{h}{\sqrt{n}} + c\frac{1}{n^K}.$$

Proof. Let $\Gamma_h(y) = e^{1/2}\exp(-y^2/2h^2)$. Then

$$\begin{aligned}\mathbb{P}^z(S_n \in x + I_h) &\leq \mathbb{E}^z\Gamma_h(S_n - x)\\ &= c\int_{\mathbb{R}} e^{-iz\theta}e^{ix\theta}\widehat{\Gamma}_h(\theta)[\varphi(\theta)]^n\,d\theta\\ &\leq \frac{c}{\sqrt{n}}\int_{\mathbb{R}}\widehat{\Gamma}_h(\alpha/\sqrt{n})|\varphi(\alpha/\sqrt{n})|^n\,d\alpha,\end{aligned}$$

where $\widehat{\Gamma}_h$ denotes the Fourier transform of Γ_h; thus $\widehat{\Gamma}_h = ch\exp(-\theta^2h^2/2) > 0$.

First,

$$\int_{|\alpha|\leq 1} \leq \frac{\|\widehat{\Gamma}_h\|_\infty}{\sqrt{n}}\int_{|\alpha|\leq 1} d\alpha \leq \frac{ch}{\sqrt{n}}.$$

As in the proof of Proposition 2.1, if r is small, $|\varphi(\alpha/\sqrt{n})|^n \leq \exp(-\alpha^2/4)$ for $1 \leq |\alpha| \leq r\sqrt{n}$, so

$$\int_{1\leq|\alpha|\leq r\sqrt{n}} \leq \frac{\|\widehat{\Gamma}_h\|_\infty}{\sqrt{n}}\int_{|\alpha|\leq r\sqrt{n}}\exp(-\alpha^2/4)\,d\alpha \leq \frac{ch}{\sqrt{n}}.$$

Finally,

$$\int_{|\alpha|\geq r\sqrt{n}} = \sqrt{n}\int_{|\theta|\geq r} \widehat{\Gamma}_h(\theta)|\varphi(\theta)|^n \, d\theta$$
$$\leq \sqrt{n}\rho_r^n \int \widehat{\Gamma}_h(\theta)\, d\theta \leq c\sqrt{n}\rho_r^n.$$

Adding these three inequalities gives our result. □

Now let $\beta \in (0,1)$ and let ψ_β be a nonnegative C^∞ function with

$$1_{[-\frac{1}{2}-\beta,\frac{1}{2}+\beta]} \geq \psi_\beta \geq 1_{[-\frac{1}{2},\frac{1}{2}]}, \quad \|\psi_\beta'\|_\infty \leq c/\beta,$$
$$\|\psi_\beta''\|_\infty \leq c/\beta^2.$$

Proposition 4.2. *Let $K \in \mathbb{Z}^+$. Then for all x, y, z*

$$|\mathbb{E}^z \psi_\beta(S_n - x) - \mathbb{E}^z \psi_\beta(S_n - y)| \leq c\left(\frac{|x-y|}{\sqrt{n}}\right)^\gamma \left(\frac{1}{\sqrt{n}} + \frac{1}{\beta n^K}\right).$$

Proof. Letting $\widehat{\psi}_\beta$ denote the Fourier transform of ψ_β, note that $\|\widehat{\psi}_\beta\|_\infty \leq \|\psi_\beta\|_1 \leq 3$ and

$$|\theta^2 \widehat{\psi}_\beta(\theta)| = |\widehat{\psi}_\beta''(\theta)| \leq \|\psi_\beta''\|_1 \leq c/\beta.$$

It follows that $|\widehat{\psi}_\beta(\theta)| \leq c/\beta\theta^2$, and so $\int_{|\theta|\geq r} |\theta|^\gamma |\widehat{\psi}_\beta(\theta)|\, d\theta \leq c/\beta$ if $\gamma \in (0,1)$, $r > 0$.

With these estimates on $\|\widehat{\psi}_\beta\|_\infty$ and $\|\widehat{\psi}_\beta\|_1$, the proof is exactly the same as the proof of Proposition 4.1, except that just as in the proof of Proposition 2.1, we use

$$|e^{ix\alpha/\sqrt{n}} - e^{iy\alpha/\sqrt{n}}| \leq |\alpha|^\gamma (|x-y|/\sqrt{n})^\gamma$$

to get the extra $(|x-y|/\sqrt{n})^\gamma$ term. □

Define

$$\eta(x,k)=\sum_{j=1}^{k}1_{[-1/2,1/2)}(S_j-x),\qquad \widetilde{\eta}_\beta(x,k)=\sum_{j=1}^{k}\psi_\beta(S_j-x),$$

and

$$L^n(x,t)=\frac{1}{\sqrt{n}}\eta(x\sqrt{n},nt),\qquad \widetilde{L}^n_\beta(x,t)=\frac{1}{\sqrt{n}}\widetilde{\eta}_\beta(x\sqrt{n},nt).$$

Proposition 4.3. *Let $\delta\in(0,1/2)$. Suppose $\beta n^{K-1}\geq 1$. Then there exist c_1,c_2 such that*

$$\mathbb{P}\Big(\sup_{x,y\in\mathbb{R},t\leq 1}|\widetilde{L}^n_\beta(x,t)-\widetilde{L}^n_\beta(y,t)|/(|x-y|^{1/2-\delta}\wedge 1)\geq\lambda/\beta\Big)$$
$$\leq c_1\exp(-c_2\lambda)+n^{-7}$$

and for large n (depending on ω)

$$\sup_{x,y\in\mathbb{R},t\leq 1}|\widetilde{L}^n_\beta(x,t)-\widetilde{L}^n_\beta(y,t)|/(|x-y|^{1/2-\delta}\wedge 1)\leq\frac{c}{\beta}\log n,\qquad a.s.$$

Proof. Let

$$A_k=\frac{1}{\sqrt{n}}(\widetilde{\eta}_\beta(x,k)-\widetilde{\eta}_\beta(y,k)).$$

By Proposition 4.2,

$$|\mathbb{E}^z A_k|\leq c\Big(\frac{|x-y|}{\sqrt{n}}\Big)^\gamma.$$

Note that if $|x-y| \geq 1$,

$$\mathbb{E}^z \sum_{j=0}^{n-1} (\Delta A_j)^2 \leq \frac{1}{n} \mathbb{E}^z \sum_{j=0}^{n-1} \Big(1_{[x-1,x+1]}(S_j) + 1_{[y-1,y+1]}(S_j) \Big)$$

$$\leq \frac{c}{\sqrt{n}} \leq \frac{c|x-y|}{\sqrt{n}\beta}$$

by using Proposition 4.1.

On the other hand, if $|x-y| \leq 1$,

$$|\psi_\beta(S_j - x) - \psi_\beta(S_j - y)| \leq \|\psi'_\beta\|_\infty |x-y| \leq c|x-y|/\beta$$

and

$$\mathbb{E}^z \sum_{j=0}^{n-1} (\Delta A_j)^2 = \frac{1}{n} \mathbb{E}^z \sum_{j=0}^{n-1} [\psi_\beta(S_j - x) - \psi_\beta(S_j - y)]^2$$

$$\leq \frac{c|x-y|}{\sqrt{n}\beta} \mathbb{E}^z \frac{1}{\sqrt{n}} \sum_{j=0}^{n-1} [\psi_\beta(S_j - x) + \psi_\beta(S_j - y)]$$

$$\leq \frac{c|x-y|}{\sqrt{n}\beta} \mathbb{E}^z \frac{1}{\sqrt{n}} \sum_{j=0}^{n-1} [1_{[x-1,x+1]}(S_j) + 1_{[y-1,y+1]}(S_j)]$$

$$\leq \frac{c|x-y|}{\sqrt{n}\beta}.$$

Similarly,

$$\sup_{j \leq n-1} |\Delta A_j|^2 \leq \frac{c|x-y|}{\beta n}.$$

With these estimates in place of (3.2), we may now proceed exactly as in Proposition 3.1 to obtain our result. □

To go from $\widetilde{L}^n_\beta(x,t)$ to $L^n(x,t)$ we have

Proposition 4.4. *If* $\beta_n \to 0$ *so that* $\liminf \log \beta_n / \log n > -\infty$*, then there exists c such that for large n (depending on* ω*)*

$$(a) \qquad \sup_{x\in\mathbb{R}, t\le 1} |\widetilde{L}^n_{\beta_n}(x,t) - L^n(x,t)| \le c\beta_n \log n, \quad a.s.$$

and

$$(b) \qquad \sup_{x\in\mathbb{R}} \frac{1}{\sqrt{n}} \sum_{j=1}^{n} 1_{[x,x+\beta_n]}(S_j) \le c\beta_n \log n, \quad a.s.$$

Proof. Choose K so that $n^{-(K-1)} \le \beta_n$. If I is an interval of length $2\beta_n$,

$$\sup_x \mathbb{P}^x(S_j \in I) \le c\beta_n/\sqrt{j} + c/j^K$$

by Proposition 4.1. Summing,

$$\sup_x \mathbb{E}^x \sum_{j=1}^{n} 1_I(S_j) \le c\beta_n\sqrt{n} + c/n^{K-1} \le 2c\beta_n\sqrt{n}.$$

The quantity $A_k = \sum_{j=1}^{k} 1_I(S_j)$ is a subadditive functional. By Chebyshev's inequality,

$$\sup_x \mathbb{P}^x(A_n > 4c\beta_n\sqrt{n}) \le 1/2.$$

So by the strong Markov property,

$$\mathbb{P}^x(A_n > 4mc\beta_n\sqrt{n}) \le (1/2)^m.$$

Therefore

$$\mathbb{P}^0\left(\frac{1}{\sqrt{n}} \sum_{j=1}^{n} 1_I(S_j) \ge c_1\beta_n \log n/2\right) \le \exp(-c_1c_2 \log n).$$

If $|\widetilde{L}^n_{\beta_n}(x,t) - L^n(x,t)| \geq c_1\beta_n \log n$, for some x, then either (i) $\sup_{j\leq n}|S_j| \geq n^4$, or (ii) for some interval I of length β_n contained in $[-n^4-1, n^4+1]$, we have $(1/\sqrt{n})\sum_{j=1}^n 1_I(S_j) \geq c_1\beta_n \log n/2$. Any interval of length β_n is contained in some interval $[k\beta_n, (k+2)\beta_n]$ for some integer k. There are at most $2n^4/\beta_n$ such intervals contained in $[-n^4-1, n^4+1]$. So the probability of possibility (ii) is bounded by $(2n^4/\beta_n)\exp(-c_1c_2\log n)$. The probability of possibility (i) is bounded by n^{-7} by Chebyshev's inequality. The result (a) follows by Borel–Cantelli if we use the assumption on β_n and take c_1 large enough. The same proof also shows (b). □

Now define the triangular arrays X_j^n, the partial sums, the local times, and the Brownian motions similarly to what was done in Section 3.

Theorem 4.5 *Let $\sigma > 0$. Suppose a_n is a sequence such that $a_n \leq (\log n)^{-10}$. Suppose*

$$\sup_{t\leq 1}|S^n_{[nt]}/\sqrt{n} - B^n_t| = O(a_n), \qquad a.s.$$

Then

$$\sup_{x\in\mathbb{R},t\leq 1}|L^n(x,t) - \ell^n(x,t)|$$
$$= O(n^{-1/(9+\sigma)} \vee a_n^{1/(9+\sigma)}\log n), \qquad a.s.$$

Proof. The proof follows the lines of the proof of Theorem 3.2 closely, but with the following changes. Let $f_{\varepsilon n}$ be the step function which takes the value $f_\varepsilon(x_0 + k/\sqrt{n})$ for

$x \in [x_0 + k/\sqrt{n}, x_0 + (k+1)/\sqrt{n})$, $k \in \mathbb{Z}$. Then

$$\left| \frac{1}{n} \sum_{j=1}^{n} f_\varepsilon(S_j/\sqrt{n}) - \frac{1}{n} \sum_{j=1}^{n} f_{\varepsilon n}(S_j/\sqrt{n}) \right| \\ \leq \|f_\varepsilon - f_{\varepsilon n}\|_\infty \leq 1/\varepsilon\sqrt{n}.$$

Note that

$$\frac{1}{n} \sum_{j=1}^{n} f_{\varepsilon n}(S_j/\sqrt{n}) = \frac{1}{n} \sum_{y \in \mathbb{Z}/\sqrt{n}} f_\varepsilon(y) \eta^n(\sqrt{n}y + \frac{1}{2}, n).$$

Also, by Propositions 4.3 and 4.4, if $|y - x_0| \leq \varepsilon$,

$$|L^n(y,1) - L^n(x_0,1)| \leq \frac{c}{\beta_n} \varepsilon^{1/2-\delta} \log n + c\beta_n \log n.$$

With these changes to (3.11), we get

$$|L^n(x_0,1) - \ell^n(x_0,1)| \leq \frac{c\varepsilon^{1/2-\delta}}{\beta_n} \log n + \frac{a_n}{\varepsilon^2} \\ + \frac{c \log n}{\varepsilon n} + \frac{1}{\varepsilon\sqrt{n}} + c\beta_n \log n.$$

Now choose $\varepsilon = \varepsilon_n = a_n^{4/9} \vee n^{-1/4}$ and $\beta_n = \varepsilon^{1/4-\delta/2}$. □

Returning to a single sequence S_n and a single Brownian motion B_t with local times $\ell(x,t)$, we also have a uniform weak invariance principle.

Theorem 4.6. *For mean 0, variance 1 strongly nonlattice random walks,*

$$\{(S_{[nt]}/\sqrt{n}, L^n(x,t)) : t \leq 1, x \in \mathbb{R}\}$$

converges weakly to

$$\{(B_t, \ell(x,t)) : t \leq 1, x \in \mathbb{R}\}.$$

Proof. One can find another probability space which contains random walks with the same distribution as S_n and a Brownian motion such that the normalized random walks converge a.s. to the Brownian motion, uniformly over $t \leq 1$. In particular, they converge in probability.

We now imitate the proofs of Theorem 4.5 and Corollary 3.3 to conclude that the $L^n(x,t)$ converge in probability to $\ell^n(x,t)$, uniformly over $t \leq 1, x \in \mathbb{R}$. This proves weak convergence. □

REFERENCES

[BK1] R.F. Bass and D. Khoshnevisan, Local times on curves and uniform invariance principles, *Probab. Th. and rel. Fields*, To appear.

[BK2] R.F. Bass and D. Khoshnevisan, Rates of convergence to Brownian local time, Preprint.

[Bo1] A.N. Borodin, On the asymptotic behavior of local times of recurrent random walks with finite variance, *Theor. Prob. and Applic.* **26**(1981), 758–772.

[Bo2] A.N. Borodin, Brownian local time, *Russian Math. Surveys* **44**(1989), 1–51.

[CH] M. Csörgö and L. Horváth, On best possible approximations of local time, *Stat. & Prob. Letters* **8**(1989), 301–306.

[DM] C. Dellacherie and P.-A. Meyer, *Probabilités et Potentiel: Théorie des Martingales*, Hermann, Paris, 1980.

[E] U. Einmahl, Strong invariance principles for partial sums of independent random vectors, *Ann. Probab.* **15**(1987), 1419–1440.

[R] P. Révész, Local times and invariance, in *Analytic Methods in Probability Theory*, LNM **861**, 128–145, Springer, Berlin, 1981.

[Sp] F. Spitzer, *Principles of Random Walk*, 2nd ed., Springer, Berlin, 1976.

[St] C.J. Stone, A local limit theorem for nonlattice multi-dimensional distribution functions, *Ann. Math. Stat.* **36** (1965), 546–551.

Department of Mathematics
University of Washington
Seattle, Washington 98195

SOME PATH PROPERTIES OF ITERATED BROWNIAN MOTION

KRZYSZTOF BURDZY

University of Washington

1. Introduction and main results. Suppose that X^1, X^2 and Y are independent standard Brownian motions starting from 0 and let

$$X(t) = \begin{cases} X^1(t) & \text{if } t \geq 0, \\ X^2(-t) & \text{if } t < 0. \end{cases}$$

We will consider the process $\{Z(t) \stackrel{\text{df}}{=} X(Y(t)), t \geq 0\}$ which we will call "iterated Brownian motion" or simply IBM. Funaki (1979) proved that a similar process is related to "squared Laplacian." Krylov (1960) and Hochberg (1978) considered finitely additive signed measures on the path space corresponding to squared Laplacian (there exists a genuine probabilistic approach, see, e.g., Mądrecki and Rybaczuk (1992)). A paper of Vervaat (1985) contains a section on the composition of self-similar processes.

The present paper is devoted to studying path properties of IBM. We want to examine how the lack of independence of increments influences the results and estimates which are well understood in the Brownian case. This may be viewed as a prelude to a deeper study of the process.

First, however, we will address the following problem.

Supported in part by NSF grant DMS 91-00244 and AMS Centennial Research Fellowship .

Problem 1. *Given* $\{Z(t), t \geq 0\}$, *can one determine* $\{X(t), t \geq 0\}$ *and* $\{Y(t), t \geq 0\}$*?*

We start with a remark that there are many examples of functions $f, g, \widetilde{f}$ and $\widetilde{g}$ such that

$$f(g(t)) = \widetilde{f}(\widetilde{g}(t)) \tag{1}$$

for all $t \geq 0$ although f and $\widetilde{f}$ are not identically equal to each other and neither are g and $\widetilde{g}$. Indeed, given any f and g, identity (1) will be satisfied if we take $\widetilde{f}(t) \stackrel{\mathrm{df}}{=} f(t/2)$ and $\widetilde{g}(t) \stackrel{\mathrm{df}}{=} 2g(t)$. This simple observation does not imply the negative answer to Problem 1 because if $\{X(t), t \geq 0\}$ and $\{Y(t), t \geq 0\}$ are "typical" Brownian paths then $\{X(t/2), t \geq 0\}$ and $\{2Y(t), t \geq 0\}$ are not typical — they have the "wrong" quadratic variation.

Suppose that processes X and Y are defined on a probability space Ω. We will write $X^\omega(t)$ and $Y^\omega(t)$ to indicate the dependence on $\omega \in \Omega$.

Theorem 1. *There is a set* $N \subset \Omega$ *of probability 0 such that if* $\omega, \omega' \in \Omega \setminus N$ *and*

$$X^\omega(Y^\omega(t)) = X^{\omega'}(Y^{\omega'}(t))$$

for all $t \geq 0$ *then either*

$$X^\omega(t) = X^{\omega'}(t),\ t \in \mathbb{R}, \quad \text{and} \quad Y^\omega(t) = Y^{\omega'}(t),\ t \geq 0,$$

or

$$X^\omega(t) = X^{\omega'}(-t),\ t \in \mathbb{R}, \quad \text{and} \quad Y^\omega(t) = -Y^{\omega'}(t),\ t \geq 0.$$

Problem 2. *Given $\{X(X(t)), t \geq 0\}$, can one determine $\{X(t), t \geq 0\}$?*

Theorem 2. *There is a set $N \subset \Omega$ of probability 0 such that if $\omega, \omega' \in \Omega \setminus N$ and*

$$X^{\omega}(X^{\omega}(t)) = X^{\omega'}(X^{\omega'}(t))$$

for all $t \geq 0$ then $X^{\omega}(t) = X^{\omega'}(t)$ for all $t \in \mathbb{R}$.

The following problem is analogous to (1). Do there exist continuous functions $f : \mathbb{R} \to \mathbb{R}$ and $\widetilde{f} : \mathbb{R} \to \mathbb{R}$ such that $f(0) = \widetilde{f}(0) = 0$ and

$$f(f(t)) = \widetilde{f}(\widetilde{f}(t)) \tag{2}$$

for all $t \geq 0$?

Example 1. We will construct functions f and $\widetilde{f}$ which satisfy (2) and "look like" Brownian paths. Choose any continuous function f which satisfies the following conditions.

(i) $f(t) = f(-t)$ for all $t \geq 0$,
(ii) $f(t) \in [0, 1]$ for all $t \in \mathbb{R}$,
(iii) $f(t) = f(1-t)$ for all $t \in [0, 1]$,
(iv) $f(0) = 0$ and $f(2) = 1/2$,
(v) $f(t) \neq 1/2$ for some $t > 2$,
(vi) f is nowhere differentiable and the quadratic variation of f over $[s, t]$ is equal to $t - s$ for all $0 < s < t < \infty$.

Then let $\widetilde{f}(t) = f(t)$ for $t \in [-2, 2]$ and $\widetilde{f}(t) = 1 - f(t)$ for $|t| \geq 2$. It is easy to see that there exist functions f and $\widetilde{f}$ satisfying all of the above conditions. It is elementary to check that functions f and $\widetilde{f}$ satisfy (2).

The above example is not totally satisfactory because it hinges on the fact that the functions f and $\widetilde{f}$ are bounded and, therefore, they lack one of the simplest properties of Brownian paths. We give a proof of Theorem 2 which uses the same rather deep properties of Brownian paths which are utilized in the proof of Theorem 1. One may ask whether Theorem 2 may be proved using only simple facts about Brownian trajectories.

Problem 3. *Suppose that X is a Brownian sheet and Y is a two-dimensional Brownian motion starting from $(2,2)$ and stopped at the hitting time T of the circle $\{z \in \mathbb{R}^2 : |z - (2,2)| = 1\}$. Let $Z(t) = X(Y(t))$. Given $\{Z(t), t \geq 0\}$ and $Y(T)$, can one determine $\{Y(t), t \geq 0\}$?*

It is quite obvious that Z cannot determine X as Y visits only a compact subset of the plane. We do not know the answer to Problem 3.

Now we turn to the Law of Iterated Logarithm. A standard proof of LIL for Brownian motion (Karatzas and Shreve (1988) 2.9.23) uses the independence of Brownian increments. Since the increments of IBM are not independent, one may wonder whether a version of the LIL holds for IBM and how difficult it is to prove. Let us start with some simple observations. By the usual LIL, a.s. for every $\varepsilon > 0$ there exists $t_0 = t_0(\varepsilon) > 0$ such that

$$|X(t)| \leq (1+\varepsilon)\sqrt{2t \log\log(1/t)}$$

and

$$|Y(t)| \leq (1+\varepsilon)\sqrt{2t \log\log(1/t)}$$

provided $|t| < t_0$. Hence for small $t > 0$,

$$\begin{aligned}
|Z(t)| &= |X(Y(t))| \\
&\leq (1+\varepsilon)\sqrt{2(1+\varepsilon)\sqrt{2t\log\log(1/t)}}\times \\
&\qquad \times \sqrt{\log\log(1/(1+\varepsilon)\sqrt{2t\log\log(1/t)})} \\
&\leq (1+\varepsilon)^2 2^{3/4} t^{1/4} (\log\log(1/t))^{3/4}.
\end{aligned}$$

Since $\varepsilon > 0$ is arbitrary, we obtain

$$\limsup_{t \to 0} \frac{|Z(t)|}{t^{1/4}(\log\log(1/t))^{3/4}} \leq 2^{3/4} \quad \text{a.s.}$$

Theorem 3 below shows that this result is not sharp. The reason is that in its derivation we have assumed the "worst case" scenario in which some unusually large increments of X match those of Y. Thus it is no surprise that the constant $2^{3/4}$ on the right hand side of the last formula is replaced with a smaller constant $2^{5/4}/3^{3/4}$ in Theorem 3.

The lower bound in the proof of LIL for Brownian motion uses the part of the Borel-Cantelli Lemma which requires independent events. It is perhaps worth noting that despite the lack of independence of increments of Z, the proof of LIL for IBM is not much harder than that for Brownian motion. This may serve as an encouragement to further study of IBM.

Theorem 3. *With probability 1,*

$$\limsup_{t\to 0} \frac{Z(t)}{t^{1/4}(\log\log(1/t))^{3/4}} = \frac{2^{5/4}}{3^{3/4}}.$$

Theorem 3 and standard arguments easily imply that

$$\limsup_{h\downarrow 0} \frac{Z(t+h)-Z(t)}{h^{1/4}(\log\log(1/h))^{3/4}} = \frac{2^{5/4}}{3^{3/4}} \quad \text{a.s.}$$

for every $t > 0$.

We are grateful to Carl Mueller and Terry McConnell for supplying references. We would also like to thank Rich Bass, Davar Khoshnevisan, Andrzej Mądrecki, Ron Pyke, Paul Tseng and Wim Vervaat for the most useful discussions. We gratefully acknowledge that the idea of Example 1 is due to an anonymous person who kindly responded to our inquiry.

2. Proofs.

Proof of Theorem 1. The set $N \subset \Omega$ will contain several families of "exceptional" paths. First of all, we will choose N so that for all $\omega \in \Omega \setminus N$ the functions $t \to X^\omega(t)$ and $t \to Y^\omega(t)$ are continuous, non-constant and non-monotone on every interval and satisfy $X^\omega(0) = Y^\omega(0) = 0$. Hence, the local maxima of X^ω are dense on the real line. We may also suppose that if X^ω attains local maxima at t_1 and t_2 then $X(t_1) \neq X(t_2)$ (in other words, all local maxima are distinct). Moreover, we can assume that for $\omega \in \Omega \setminus N$

$$\limsup_{t\to\infty} Y^\omega(t) = -\liminf_{t\to\infty} Y^\omega(t) = \infty.$$

Suppose that $-\infty < a < b < \infty$ and let $M = \max_{t\in[a,b]} Y(t)$. Then with probability 1, $X(M)$ is not a maximum of X over any interval $[M-\varepsilon, M]$ or $[M, M+\varepsilon]$ for any $\varepsilon > 0$. The same is true if we let $M = \min_{t\in[a,b]} Y(t)$. By taking the union over all rational pairs (a,b) we may choose N so that for $\omega \in \Omega \setminus N$ and every local extremum M of Y^ω, $X^\omega(M)$ is not a maximum of X^ω over any interval $[M-\varepsilon, M]$ or $[M, M+\varepsilon]$ for any $\varepsilon > 0$. Let us call this property (P). This completes our description of N for now. We will throw in some more sets into N later.

We will argue that if M is a local maximum of Z then it is also a local maximum of X. Suppose that $Z(t) = X(Y(t)) = M$ is a local maximum of Z and let $s = Y(t)$. Assume that $M = X(s)$ is not a local maximum for X. Then for every $\varepsilon > 0$ the process X takes a value greater than M in $[s-\varepsilon, s]$ or in $[s, s+\varepsilon]$ and without loss of generality we may assume that for every $\varepsilon > 0$ we have $X(v) > M$ for some $v \in [s, s+\varepsilon]$. This and the fact that $M = Z(t)$ is a local maximum for Z imply that there is some $\delta > 0$ such that $Y(u) \le s$ for all $u \in [t-\delta, t+\delta]$. In other words, $s = Y(t)$ is a local maximum for Y. Moreover, we must have $X(v) \le M$ for all $v \in [s-\delta_1, s]$ for some $\delta_1 > 0$. Hence, M is the maximum of X over $[s-\delta_1, s]$. This and the fact that $s = Y(t)$ is a local maximum for Y contradict property (P). We conclude that if M is a local maximum of Z then it must be also a local maximum of X.

Vice versa, if M is a local maximum of X then it is also a local maximum of Z. Suppose that $M = X(t)$ is the maximum of X over $(t-\varepsilon, t+\varepsilon)$ for some $\varepsilon > 0$. Find s and $\delta > 0$ such that $Y(s) = t$ and $Y(u) \in (t-\varepsilon, t+\varepsilon)$ for $u \in (s-\delta, s+\delta)$. Then clearly M is the maximum of Z over $(s-\delta, s+\delta)$.

Let us list some elementary properties of IBM $Z(t)$ which hold for all $\omega \in \Omega \setminus N$. We have $Z(0) = 0$, the function $t \to Z(t)$ is continuous, non-constant and non-monotone on every interval. Hence, the set of times when Z attains a local maximum is dense in $[0, \infty)$.

We remark parenthetically that if $M = Z(t)$ is a local maximum of Z then for each $\varepsilon > 0$ the interval $[t-\varepsilon, t+\varepsilon]$ contains uncountably many s such that $Z(s) = M$. We skip an easy proof as we do not need this fact in the sequel. We have to remember though that two local maxima of Z attained at different points may have the same value.

Let $\{M_1, M_2, \dots\}$ be the sequence of the values of all local maxima of Z ordered in an arbitrary way. We do not place the same value more than once in the sequence. In other words, $M_j \neq M_k$ if $j \neq k$. For $j \geq 1$, let s_j be such that X attains the local maximum M_j at s_j. Recall that s_j is uniquely defined. The local maxima M_1, M_2 and M_3 come in a certain order, say, $s_1 < s_3 < s_2$. We may gain some knowledge about the order in which they come by examining the path of Z as follows. There exist $t_1 < t_3 < t_2$ such that $Y(t_j) = s_j$ for $j = 1, 2, 3$ and $Y(t) \in (s_1, s_2)$ for all $t \in (t_1, t_2)$. It follows easily that Z attains the local maximum M_j at t_j for $j = 1, 2, 3$, and, moreover, if Z attains a local maximum at $t \in (t_1, t_3)$ then this maximum cannot be equal to M_2. Likewise, any local maximum of Z between t_3 and t_2 cannot be equal to M_1. It is easy to see that Z may have the last two properties only if s_3 is between s_1 and s_2 (although the precise order may be $s_1 < s_3 < s_2$ or $s_2 < s_3 < s_1$). Our argument can be repeated for any triplet $(M_{j_1}, M_{j_2}, M_{j_3})$ of local maxima of X.

We will build a function $\{\widetilde{V}(t), t \in (-\infty, \infty)\}$ and later "time-change" $\widetilde{V}$ to obtain $\{X(t), t \in (-\infty, \infty)\}$. First we will inductively define a function $V(u)$ for certain values of u. We start with $V(1) \stackrel{\text{df}}{=} M_1$ and $V(2) \stackrel{\text{df}}{=} M_2$. Suppose that the function V is defined at points u_j, $1 \leq j \leq k$, and $V(u_j) = M_j$ for $j \leq k$. Now we use the argument from the previous paragraph to determine for each triplet $(M_{j_1}, M_{j_2}, M_{j_3})$, $j_1, j_2, j_3 \le k+1$, of local maxima which maximum comes between the other two on the path of X. The set $A_k \stackrel{\mathrm{df}}{=} \mathbb{R} \setminus \bigcup_{j \le k}\{u_j\}$ consists of a finite number of intervals and we will choose u_{k+1} in one of the intervals (say, I) so that for each triplet $(u_{j_1}, u_{j_2}, u_{j_3})$, $j_1, j_2, j_3 \le k+1$, the point u_{j_1} is between u_{j_2} and u_{j_3} if and only if s_{j_1} is between s_{j_2} and s_{j_3} on the path of X. It is easy to see that such a choice of I is possible and unique. To be more specific, if the endpoints of I are u_{j_1} and u_{j_2} then we let $u_{k+1} = (u_{j_1} + u_{j_2})/2$. If the endpoints of I are u_{j_1} and ∞ then $u_{k+1} = u_{j_1} + 1$ and finally, if the endpoints are $-\infty$ and u_{j_1} then $u_{k+1} = u_{j_1} - 1$. Then we let $V(u_{k+1}) = M_{k+1}$.

We will show that the function $V(u)$ is defined for a set of u's which is dense in $\mathbb{R}$. If there is $\widetilde{u} < \infty$ such that our procedure does not define $V(u)$ for $u > \widetilde{u}$ then there must exist the greatest or the smallest time when the Brownian motion X attains a local maximum — this is impossible. For the same reason there is no finite lower bound for the set of u's for which V is defined. Suppose that there exists an open interval (v_1, v_2) such that $V(u)$ is not defined for any $u \in (v_1, v_2)$. We may suppose that (v_1, v_2) is a maximal interval with this property and we see that because of the way we choose u_j's, we must have $v_1 = u_{j_1}$ and $v_2 = u_{j_2}$ for some j_1 and j_2. This implies that there is no maximum on the path of X between s_{j_1} and s_{j_2}. This cannot happen and our assertion is proved.

Next we will show that V can be continuously extended to $\mathbb{R}$. Recall that $X(s_j) = V(u_j) = M_j$. Let us define a function λ on $\bigcup_j\{s_j\}$ by $\lambda(s_j) \stackrel{\mathrm{df}}{=} u_j$. It follows from our construction of V that the function λ is strictly monotone. Moreover it cannot have jumps because then $\bigcup_j\{u_j\}$ could not be dense in $\mathbb{R}$. Hence, λ and λ^{-1} are well-defined, strictly monotone and continuous. Let $\widetilde{V}(u) \stackrel{\mathrm{df}}{=} X(\lambda^{-1}(u))$ for all real u. This function is a.s. continuous since it is a composition of two continuous functions. The functions V and $\widetilde{V}$ agree on a dense subset of the real line, namely on $\bigcup_j\{u_j\}$. We conclude that $\widetilde{V}$ is a (necessarily unique) continuous extension of V to the whole real line.

We have proved that Z^ω uniquely determines a continuous function $\widetilde{V}^\omega$ which is a (random) time-change of X^ω by a strictly monotone and continuous function λ_ω^{-1}. Suppose that there exist ω and ω' such that

$$\widetilde{V}^\omega(u) = X^\omega(\lambda_\omega^{-1}(u)) = X^{\omega'}(\lambda_{\omega'}^{-1}(u))$$

for all u. Then there exists a strictly monotone and continuous function $\gamma = \gamma_{\omega,\omega'}$ such that $X^\omega(t) = X^{\omega'}(\gamma(t))$ for all t. Theorem 1.5.8 of Karatzas and Shreve (1988) easily implies the following two facts. First, with probability 1, for all rational s and t, the quadratic variation of X over $[s,t]$ is equal to $t-s$ (we add the trajectories which do not have this property to the set N). Second, for all rational s and t, the quadratic variation of X^ω over $[s,t]$ is equal to the quadratic variation of $X^{\omega'}$ over $[\gamma(s),\gamma(t)]$ assuming that $\omega,\omega' \notin N$ (note that γ maps a sequence of partitions of $[s,t]$ with the mesh going to 0 onto a sequence of partitions of $[\gamma(s),\gamma(t)]$ with the mesh going to 0). The last two facts imply that γ is a linear function with the slope equal to 1 or -1. It follows that λ^{-1} is unique up to an additive constant and up to the multiplication by -1. We can find a sequence of local maxima $\{K_j\}_{j\geq 0}$ of Z such that $Z(t_j) = K_j$ and $t_j \to 0$ as $j \to \infty$. Let s_j be such that $\widetilde{V}(s_j) = K_j$ and $\widetilde{V}$ attains a local maximum at s_j. The s_j's are uniquely determined and they must converge to a point which corresponds to the origin on the path of X. More precisely, s_j's converge to s such that $\lambda^{-1}(s) = 0$. This determines λ^{-1} up to the multiplication by -1. Hence we either have $X^\omega(t) = X^{\omega'}(t)$ for all $t \in \mathbb{R}$ or $X^\omega(t) = X^{\omega'}(-t)$ for all $t \in \mathbb{R}$. The part of Theorem 1 concerned with the path of X is proved.

Suppose that the paths of Z and X are given. If Z attains a local maximum M at t then there exists a unique s such that $X(s) = M$ and X attains a local maximum at s. We must have $Y(t) = s$. Since the set of points where Z attains a local maximum is dense on the real line, the path of Y is determined on a dense set and, by continuity, on the whole real line. □

Proof of Theorem 2. First we prove that the paths of X have a property analogous to property (P) introduced in the proof of Theorem 1. Suppose that $-\infty < a < b < c < d < \infty$ and let $M = \max_{t\in[a,b]} X(t)$. Given the values of $X(b)$ and $X(c)$, the processes $\{X(t), t \in [a,b]\}$ and $\{X(t), t \in [c,d]\}$ are independent. Hence, by conditioning on $X(b)$ and $X(c)$, one can easily prove that with probability 1, $X(M)$ is not a maximum of X over any interval $[M-\varepsilon, M] \cap [c,d]$ or $[M, M+\varepsilon] \cap [c,d]$ for any $\varepsilon > 0$. The same is true if we let $M = \min_{t\in[a,b]} X(t)$. By taking the union over all rational numbers $a < b < c < d$ and $c < d < a < b$ we see that with probability 1, if M is a local extremum of X and $X(M) \neq M$ then $X(M)$ is not a maximum of X over any interval $[M-\varepsilon, M]$ or $[M, M+\varepsilon]$ for any $\varepsilon > 0$. For any fixed interval $[a,b]$ with rational endpoints, the maximum of X over $[a,b]$ is not attained at a time M such that $X(M) = M$, with probability 1. By taking the union over all such intervals, we see that with probability 1, $X(M) \neq M$ for every local maximum of X. We conclude that almost all trajectories of X have the property analogous to (P).

Now we can follow the proof of Theorem 1 and arrive at the conclusion that for $\omega, \omega' \in \Omega \setminus N$ the condition

$$X^\omega(X^\omega(t)) = X^{\omega'}(X^{\omega'}(t)) \quad \text{for} \quad t \geq 0,$$

implies that either

$$X^\omega(t) = X^{\omega'}(t), \ t \in \mathbb{R},$$

or

$$X^\omega(t) = X^{\omega'}(-t), \ t \in \mathbb{R}, \quad \text{and} \quad X^\omega(t) = -X^{\omega'}(t), \ t \geq 0.$$

In the latter case we would have $X^\omega(t) = -X^\omega(-t)$ for $t \geq 0$. Such paths form an event of probability 0. □

Lemma 1. *There exists $c_1 < \infty$ such that for all $t, a > 0$*

$$P(\sup_{0 \le s \le t} |Z(s)| > a) \le c_1 \exp(-3 \cdot 2^{-5/3} a^{4/3} t^{-1/3}).$$

Proof. Recall that for $a > 0$ (cf. Karatzas and Shreve (1988) 2.9.22)

$$\int_y^\infty \frac{1}{\sqrt{2\pi t}} \exp(-x^2/2t)dx \le \frac{1}{y}\sqrt{\frac{t}{2\pi}} \exp(-y^2/2t). \tag{3}$$

We derive two related estimates.

$$\begin{aligned}\int_b^\infty \sqrt{\frac{x}{t}} \exp(-x^2/2t)dx &\le \int_b^\infty \sqrt{\frac{x}{t}}\sqrt{\frac{t}{b}}\sqrt{\frac{x}{t}} \exp(-x^2/2t)dx \\ &= -\sqrt{\frac{t}{b}} \exp(-x^2/2t)\Big|_{x=b}^{x=\infty} \\ &= \sqrt{\frac{t}{b}} \exp(-b^2/2t).\end{aligned} \tag{4}$$

$$\begin{aligned}\int_0^b \sqrt{y} \exp(-a^2/2y)dy &\le \int_0^b \sqrt{b}\frac{2}{a^2}\frac{a^2}{2}\frac{b^2}{y^2} \exp(-a^2/2y)dy \\ &= \sqrt{b}\frac{2}{a^2} b^2 \exp(-a^2/2y)\Big|_{y=0}^{y=b} \\ &= 2b^{5/2}a^{-2} \exp(-a^2/2b).\end{aligned} \tag{5}$$

It is known (see Karatzas and Shreve (1988) 2.8.2) that the density of the random variable $\sup_{0 \le s \le t} Y(s)$ is equal to

$$\sqrt{2/\pi t} \exp(-y^2/2t)$$

for $y > 0$. It follows that the density of $\sup_{0\le s\le t}|Y(s)|$ is bounded by

$$2\sqrt{2/\pi t}\exp(-y^2/2t) \tag{6}$$

for $y > 0$. A similar formula holds for the density of the supremum of X. We may combine it with (3) to obtain

$$\begin{aligned} P(\sup_{0\le s\le y} X(s) > a) &= \int_a^\infty \sqrt{2/\pi y}\exp(-z^2/2y)dz \\ &\le (1/a)\sqrt{2y/\pi}\exp(-a^2/2y). \end{aligned}$$

Hence, by symmetry,

$$P(\sup_{-y\le s\le y}|X(s)| > a) \le (4/a)\sqrt{2y/\pi}\exp(-a^2/2y). \tag{7}$$

If $\sup_{0\le s\le t}|Y(s)| = y$ and $\sup_{0\le s\le t}|Z(s)| > a$ then $\sup_{-y\le s\le y}|X(s)| > a$. Thus, by conditioning on the value of $\sup_{0\le s\le t}|Y(s)|$ and using (6) and (7) we obtain

$$\begin{aligned} &P(\sup_{0\le s\le t}|Z(s)| > a) \\ &\le \int_0^\infty (4/a)\sqrt{2y/\pi}\exp(-a^2/2y)2\sqrt{2/\pi t}\exp(-y^2/2t)dy \\ &= \frac{16}{\pi a\sqrt{t}}\int_0^\infty \sqrt{y}\exp(-a^2/2y - y^2/2t)dy. \end{aligned} \tag{8}$$

We will split the interval of integration in the last formula into three subintervals and estimate the integral for each one of them. It is elementary to check that the maximum of the function

$$y \to \exp(-a^2/2y - y^2/2t)$$

on the interval $[0, \infty)$ is attained at $u \stackrel{\text{df}}{=} (ta^2/2)^{1/3}$ and so

$$\begin{aligned} \exp(-a^2/2y - y^2/2t) &\le \exp(-a^2/2u - u^2/2t) \\ &= \exp(-3 \cdot 2^{-5/3} a^{4/3} t^{-1/3}) \end{aligned}$$

for all $y > 0$. This implies that

$$\begin{aligned} &\frac{16}{\pi a \sqrt{t}} \int_{2u/3}^{\sqrt{3}u} \sqrt{y} \exp(-a^2/2y - y^2/2t) dy \\ &\quad \le \frac{16}{\pi a \sqrt{t}} \sqrt{\sqrt{3}u} \exp(-a^2/2u - u^2/2t) u(\sqrt{3} - 2/3) \\ &\quad = c_2 \frac{u^{3/2}}{a\sqrt{t}} \exp(-3 \cdot 2^{-5/3} a^{4/3} t^{-1/3}) \\ &\quad = c_2 \frac{(ta^2/2)^{1/2}}{a\sqrt{t}} \exp(-3 \cdot 2^{-5/3} a^{4/3} t^{-1/3}) \\ &\quad = c_3 \exp(-3 \cdot 2^{-5/3} a^{4/3} t^{-1/3}). \end{aligned} \tag{9}$$

Now we use (4) to obtain

$$\begin{aligned} &\frac{16}{\pi a \sqrt{t}} \int_{\sqrt{3}u}^{\infty} \sqrt{y} \exp(-a^2/2y - y^2/2t) dy \\ &\quad \le \frac{16}{\pi a} \int_{\sqrt{3}u}^{\infty} \sqrt{\frac{y}{t}} \exp(-y^2/2t) dy \\ &\quad \le \frac{c_4}{a} \sqrt{\frac{t}{\sqrt{3}u}} \exp(-(\sqrt{3}u)^2/2t) \\ &\quad = c_5 t^{1/3} a^{-4/3} \exp(-3 \cdot 2^{-5/3} a^{4/3} t^{-1/3}). \end{aligned} \tag{10}$$

The inequality (5) gives

$$\begin{aligned}\frac{16}{\pi a\sqrt{t}}\int_0^{2u/3}&\sqrt{y}\exp(-a^2/2y-y^2/2t)dy\\ &\le\frac{16}{\pi a\sqrt{t}}\int_0^{2u/3}\sqrt{y}\exp(-a^2/2y)dy\\ &\le\frac{c_6}{a\sqrt{t}}2a^{-2}(2u/3)^{5/2}\exp(-a^2/(2(2u/3)))\\ &=c_7t^{1/3}a^{-4/3}\exp(-3\cdot 2^{-5/3}a^{4/3}t^{-1/3}).\end{aligned}$$

This, (8), (9) and (10) imply that

$$\begin{aligned}P(\sup_{0\le s\le t}&|Z(s)|>a)\\ &\le(c_3+c_8t^{1/3}a^{-4/3})\exp(-3\cdot 2^{-5/3}a^{4/3}t^{-1/3}).\end{aligned}$$

Choose $c_1<\infty$ so large that $c_3+c_8x^{-1}\le c_1$ for $x>1$ and $c_1\exp(-3\cdot 2^{-5/3}x)\ge 1$ for $x\le 1$. Then

$$P(\sup_{0\le s\le t}|Z(s)|>a)\le c_1\exp(-3\cdot 2^{-5/3}a^{4/3}t^{-1/3}).\quad\square$$

Lemma 2. *Fix some $\gamma,\beta\in(0,1)$. Then for sufficiently small $u>0$ we have*

$$\begin{aligned}P(\inf_{s\in[u,3u]}X(s)-X(\gamma u)\ge(1-2\beta)(2/3^{1/2})(u\log\log(1/u))^{1/2})\\ \ge\log(1/u)^{-(2/3)(1-\beta)/(1-\gamma)}.\end{aligned}$$

Proof. If $y>\sqrt{t}$ then (cf. Karatzas and Shreve (1988) 2.9.22)

$$(11)\qquad\int_y^\infty\frac{1}{\sqrt{2\pi t}}\exp(-x^2/2t)dx\ge\frac{1}{2y}\sqrt{\frac{t}{2\pi}}\exp(-y^2/2t).$$

Thus

$$
\begin{aligned}
&P(X(u) - X(\gamma u) \geq (1-\beta)(2/3^{1/2})(u \log\log(1/u))^{1/2}) \\
&\geq \frac{((1-\gamma)u)^{1/2} \exp(-(1-\beta)^2(4/3)u \log\log(1/u)/2u(1-\gamma))}{2\sqrt{2\pi}(1-\beta)(2/3^{1/2})(u \log\log(1/u))^{1/2}} \\
&= \frac{(1-\gamma)^{1/2} \exp(-(1-\beta)^2(1-\gamma)^{-1}(2/3) \log\log(1/u))}{4 \cdot 3^{-1/2}\sqrt{2\pi}(1-\beta)(\log\log(1/u))^{1/2}} \\
&= \frac{(1-\gamma)^{1/2} \log(1/u)^{-(2/3)(1-\beta)^2/(1-\gamma)}}{4 \cdot 3^{-1/2}\sqrt{2\pi}(1-\beta)(\log\log(1/u))^{1/2}}.
\end{aligned}
$$

For small $u > 0$

$$
\begin{aligned}
P(\inf_{s\in[u,3u]} X(s) - X(u) &\geq -\beta(2/3^{1/2})(u \log\log(1/u))^{1/2}) \\
&\geq P(\inf_{s\in[u,3u]} X(s) - X(u) \geq -3\sqrt{2u}) \geq 1/2.
\end{aligned}
$$

The Markov property applied at u yields for small $u > 0$

$$
\begin{aligned}
P(\inf_{s\in[u,3u]} X(s) - X(\gamma u) &\geq (1-2\beta)(2/3^{1/2})(u \log\log(1/u))^{1/2}) \\
&\geq \frac{1}{2} \cdot \frac{(1-\gamma)^{1/2} \log(1/u)^{-(2/3)(1-\beta)^2/(1-\gamma)}}{4 \cdot 3^{-1/2}\sqrt{2\pi}(1-\beta)(\log\log(1/u))^{1/2}} \\
&\geq \log(1/u)^{-(2/3)(1-\beta)/(1-\gamma)}. \quad \square
\end{aligned}
$$

Proof of Theorem 3. First we will derive the upper bound. Let

$$
f(t) \stackrel{\text{df}}{=} (2^{5/4}/3^{3/4})t^{1/4}(\log\log(1/t))^{3/4}.
$$

Choose some $\eta > 1$. Let $\gamma = \eta^{4/3}$ and $a = \eta f(t)$. With this choice of a, Lemma 1 gives

$$
\begin{aligned}
P(\sup_{0\leq s\leq t} |Z(s)| > \eta f(t)) &= P(\sup_{0\leq s\leq t} |Z(s)| > a) \\
&\leq c_1 \exp(-3 \cdot 2^{-5/3} a^{4/3} t^{-1/3}) \\
&= c_1 \exp(-\eta^{4/3} \log\log(1/t)) \\
&= c_1(\log(1/t))^{-\gamma}.
\end{aligned}
$$

Choose an arbitrary $\alpha < 1$ and note that $\eta^2\alpha^{-1/4}f(s) > \eta f(\alpha^k)$ for $s \in [\alpha^{k+1}, \alpha^k]$ provided k is large. We apply the last formula with $t = \alpha^k$ to obtain for large k

$$\begin{aligned} P(\sup_{\alpha^{k+1}\le s\le\alpha^k} |Z(s)|/f(s) > \eta^2\alpha^{-1/4}) &\le P(\sup_{\alpha^{k+1}\le s\le\alpha^k} |Z(s)| > \eta f(\alpha^k)) \\ &\le P(\sup_{0\le s\le\alpha^k} |Z(s)| > \eta f(\alpha^k)) \\ &\le c_1(k\log(1/\alpha))^{-\gamma}. \end{aligned}$$

We have $\sum_k c_1(k\log(1/\alpha))^{-\gamma} < \infty$ since $\gamma > 1$. The Borel-Cantelli lemma implies that only a finite number of events

$$\{\sup_{\alpha^{k+1}\le s\le\alpha^k} |Z(s)|/f(s) > \eta^2\alpha^{-1/4}\}$$

occur. It follows that a.s.

$$\begin{aligned} \limsup_{t\to 0} \frac{Z(t)}{(2^{5/4}/3^{3/4})t^{1/4}(\log\log(1/t))^{3/4}} &= \limsup_{t\to 0} Z(t)/f(t) \\ &\le \eta^2\alpha^{-1/4}. \end{aligned}$$

Since η and α may be chosen arbitrarily close to 1,

$$\limsup_{t\to 0} \frac{Z(t)}{(2^{5/4}/3^{3/4})t^{1/4}(\log\log(1/t))^{3/4}} \le 1 \quad \text{a.s.} \tag{12}$$

Next we will prove the lower bound.

Fix an arbitrarily small $\varepsilon_0 > 0$ and find $\varepsilon, \alpha, \beta \in (0, \varepsilon_0)$ which satisfy the following conditions.

(i) $\gamma \overset{\text{df}}{=} 4\alpha^{1/2}/(1-2\varepsilon) < 1 - 2\varepsilon$,

(ii) $\delta \overset{\text{df}}{=} (1-\beta)/(1-\gamma) < 1$,

(iii) $(4\alpha)^{1/2} \le \varepsilon((2/3)(1-\alpha))^{1/2}/2$,
(iv) $(4 \cdot 4\alpha^{1/2})^{1/2} \le (\varepsilon/2)(1-2\beta)(2/3^{1/2})(1-2\varepsilon)^{1/2}$,
(v) $\varepsilon < 1/2$.

Let $t_k = \alpha^k$ and

$$\begin{aligned} u_k &= ((2/3)(t_k - t_{k-1}) \log\log(1/(t_k - t_{k-1})))^{1/2} \\ &= ((2/3)(1-\alpha)t_k \log\log(1/((1-\alpha)t_k)))^{1/2}. \end{aligned}$$

Note that for large k we have $3(1-2\varepsilon)u_{k+1} < 4\alpha^{1/2}u_k$.

For large k we obtain using (3) and (11)

$$\begin{aligned} &P(Y(t_k) - Y(t_{k+1}) \in [(1-\varepsilon)u_k, 2(1-\varepsilon)u_k]) \\ &= \int_{(1-\varepsilon)u_k}^{2(1-\varepsilon)u_k} \frac{1}{\sqrt{2\pi(1-\alpha)t_k}} \exp(-x^2/2(1-\alpha)t_k)dx \\ &\ge \frac{1}{2(1-\varepsilon)u_k}\sqrt{\frac{(1-\alpha)t_k}{2\pi}} \exp(-((1-\varepsilon)u_k)^2/2(1-\alpha)t_k) \\ &\quad - \frac{1}{2(1-\varepsilon)u_k}\sqrt{\frac{(1-\alpha)t_k}{2\pi}} \exp(-(2(1-\varepsilon)u_k)^2/2(1-\alpha)t_k) \\ &\ge \frac{1}{4(1-\varepsilon)u_k}\sqrt{\frac{(1-\alpha)t_k}{2\pi}} \exp(-((1-\varepsilon)u_k)^2/2(1-\alpha)t_k) \\ &= \frac{\exp(-(1-\varepsilon)^2(1/3)\log\log(1/((1-\alpha)t_k)))}{4\sqrt{2\pi}(1-\varepsilon)((2/3)\log\log(1/((1-\alpha)t_k)))^{1/2}} \\ &= \frac{(\log(1/((1-\alpha)t_k)))^{-(1-\varepsilon)^2/3}}{8(1-\varepsilon)((\pi/3)\log\log(1/((1-\alpha)t_k)))^{1/2}} \\ &\ge c_4(\log(1/((1-\alpha)t_k)))^{-(1-\varepsilon)/3} \\ &= c_4(\log(1/((1-\alpha)\alpha^k)))^{-(1-\varepsilon)/3} \end{aligned}$$

(13)

$$\ge c_5 k^{-(1-\varepsilon)/3}.$$

Let

$$g(u) \stackrel{\text{df}}{=} (1-2\beta)(2/3^{1/2})((1-2\varepsilon)u\log\log(1/(1-2\varepsilon)u))^{1/2}.$$

Since $\gamma \stackrel{\text{df}}{=} 4\alpha^{1/2}/(1-2\varepsilon) < 1-2\varepsilon$, Lemma 2 implies for large k

$$\begin{aligned}
&P\Big(\inf_{s\in[(1-2\varepsilon)u_k,3(1-2\varepsilon)u_k]} X(s) - X(4\alpha^{1/2}u_k) \geq g(u_k)\Big) \\
&\quad \geq \log(1/((1-2\varepsilon)u_k))^{-(2/3)(1-\beta)/(1-\gamma)} \\
&\quad = \log(1/((1-2\varepsilon)(\tfrac{2}{3}(1-\alpha)t_k \log\log(1/((1-\alpha)t_k)))^{\frac{1}{2}}))^{\frac{-2\delta}{3}} \\
&\quad \geq \log(1/((1-\varepsilon)((2/3)(1-\alpha)t_k)^{1/2}))^{-(2/3)\delta} \\
&\quad = \log(1/((1-\varepsilon)((2/3)(1-\alpha)\alpha^k)^{1/2}))^{-(2/3)\delta} \\
(14)\quad &\quad \geq c_6 k^{-(2/3)\delta}.
\end{aligned}$$

Let

$$\begin{aligned}
A_k \stackrel{\text{df}}{=} &\{\inf_{s\in[(1-2\varepsilon)u_k,3(1-2\varepsilon)u_k]} X(s) - X(4\alpha^{1/2}u_k) \geq g(u_k)\} \\
&\cap\{Y(t_k) - Y(t_{k+1}) \in [(1-\varepsilon)u_k, 2(1-\varepsilon)u_k]\}.
\end{aligned}$$

By (13), (14) and the independence of X and Y,

$$P(A_k) \geq c_5 k^{-(1-\varepsilon)/3} c_6 k^{-(2/3)\delta}.$$

Since $-(1-\varepsilon)/3 - (2/3)\delta > -1$,

$$\sum_k c_5 k^{-(1-\varepsilon)/3} c_6 k^{-(2/3)\delta} = \infty.$$

Recall that $3(1-2\varepsilon)u_{k+1} < 4\alpha^{1/2}u_k$ for large k. It follows that for some large k_0, the events $\{A_k\}_{k\geq k_0}$ are independent. Thus the Borel-Cantelli Lemma implies that with probability 1 infinitely many events A_k occur.

We have chosen α and ε so that

$$(4\alpha)^{1/2} \leq \varepsilon((2/3)(1-\alpha))^{1/2}/2.$$

The standard LIL implies that with probability 1 for all sufficiently large k we have

$$\begin{aligned} |Y(t_{k+1})| &\leq (4t_{k+1} \log\log(1/t_{k+1}))^{1/2} \\ &= (4\alpha t_k \log\log(1/\alpha t_k))^{1/2} \\ &\leq \varepsilon((2/3)(1-\alpha)t_k \log\log(1/((1-\alpha)t_k)))^{1/2} \\ &= \varepsilon u_k. \end{aligned} \tag{15}$$

We have assumed that

$$(4 \cdot 4\alpha^{1/2})^{1/2} \leq (\varepsilon/2)(1-2\beta)(2/3^{1/2})(1-2\varepsilon)^{1/2}.$$

The LIL for the Brownian motion X shows that for large k

$$\begin{aligned} &|X(4\alpha^{1/2}u_k)| \\ &\leq (4 \cdot 4\alpha^{1/2}u_k \log\log(1/(4\alpha^{1/2}u_k)))^{1/2} \\ &\leq (\varepsilon/2)(1-2\beta)(2/3^{1/2})((1-2\varepsilon)u_k \log\log(1/(4\alpha^{1/2}u_k)))^{1/2} \\ &\leq \varepsilon(1-2\beta)(2/3^{1/2})((1-2\varepsilon)u_k \log\log(1/(1-2\varepsilon)u_k))^{1/2} \\ &= \varepsilon g(u_k). \end{aligned} \tag{16}$$

Recall that infinitely many events

$$\begin{aligned} A_k = &\{\inf_{s\in[(1-2\varepsilon)u_k, 3(1-2\varepsilon)u_k]} X(s) - X(4\alpha^{1/2}u_k) \geq g(u_k)\} \\ &\cap \{Y(t_k) - Y(t_{k+1}) \in [(1-\varepsilon)u_k, 2(1-\varepsilon)u_k]\} \end{aligned}$$

occur a.s. By (15) and (16), the events $\{|Y(t_{k+1})| \le \varepsilon u_k\}$ and $\{|X(4\alpha^{1/2}u_k)| \le \varepsilon g(u_k)\}$ hold for all large k a.s. Hence

$$Y(t_k) \in [(1-2\varepsilon)u_k, 3(1-2\varepsilon)u_k]$$

and

$$\inf_{s\in[(1-2\varepsilon)u_k,3(1-2\varepsilon)u_k]} X(s) \ge (1-\varepsilon)g(u_k)$$

for infinitely many k a.s. It follows that $Z(t_k) \ge (1-\varepsilon)g(u_k)$ occurs i.o. with probability 1. It is easy to check that

$$\frac{(1-2\beta)(2/3^{1/2})((1-2\varepsilon)u_k \log\log(1/(1-2\varepsilon)u_k))^{1/2}}{(1-2\beta)(2/3^{1/2})((1-2\varepsilon)((2/3)(1-\alpha)t_k)^{1/4}(\log\log(1/t_k))^{3/4}}$$

goes to 1 as $k \to \infty$. Hence

$$\begin{aligned}
&\limsup_{k\to\infty} \frac{Z(t_k)}{(2^{5/4}/3^{3/4})t_k^{1/4}(\log\log(1/t_k))^{3/4}} \times \\
&\qquad \times \frac{1}{(1-2\beta)(1-2\varepsilon)^{1/4}(1-\alpha)^{1/4}} \\
&= \limsup_{k\to\infty} \frac{Z(t_k)}{\frac{2(1-2\beta)}{\sqrt{3}}((1-2\varepsilon)(\frac{2}{3}(1-\alpha)t_k)^{1/4}(\log\log(1/t_k))^{3/4}} \\
&= \limsup_{k\to\infty} \frac{Z(t_k)}{(1-2\beta)\frac{2}{\sqrt{3}}((1-2\varepsilon)u_k \log\log(1/(1-2\varepsilon)u_k))^{1/2}} \\
&= \limsup_{k\to\infty} \frac{Z(t_k)}{g(u_k)} \\
&\ge (1-\varepsilon) \quad \text{a.s.}
\end{aligned}$$

Since ε, α and β may be chosen arbitrarily close to 0, we obtain

$$\limsup_{t\to 0} \frac{Z(t)}{(2^{5/4}/3^{3/4})t^{1/4}(\log\log(1/t))^{3/4}} \ge 1 \quad \text{a.s.} \tag{17}$$

This and (12) prove Theorem 3. □

REFERENCES

1. T. Funaki, *Probabilistic construction of the solution of some higher order parabolic differential equations*, Proc. Japan Acad. **55** (1979), 176–179.
2. K.J. Hochberg, *A signed measure on path space related to Wiener measure*, Ann. Probab. **6** (1978), 433–458.
3. I. Karatzas and S. Shreve, *Brownian Motion and Stochastic Calculus*, Springer, New York, 1988.
4. V. Yu. Krylov, *Some properties of the distribution corresponding to the equation* $\partial u/\partial t = (-1)^{q+1}\partial^{2q}u/\partial x^{2q}$, Soviet Math. Dokl. **1** (1960), 760–763.
5. A. Mądrecki and M. Rybaczuk, *New Feynman-Kac type formula*, (preprint) (1992).
6. W. Vervaat, *Sample path properties of self-similar processes with stationary increments*, Ann. Probab. **13** (1985), 1–27.

DEPARTMENT OF MATHEMATICS, GN-50 SEATTLE, WA 98195

Conditioned superprocesses and a semilinear heat equation

Alison M. Etheridge

1. Introduction

Measure-valued branching processes (hereafter called MVB processes and denoted X_t) have been the subject of intensive study over the last twenty years. We recommend Dawson [2] as an excellent general reference and for the rest of this note assume some familiarity with the construction and characterisation of these processes. We shall be concerned only with the special case when the branching mechanism has finite variance and the MVB process takes its values in $\mathcal{M}_F(E)$ (finite measures on a Polish space E). In particular we will be motivated by a pathwise construction of the MVB process due to Le Gall [9].

We will assume that the spatial motion underlying the MVB process is a conservative Feller process with infinitesimal generator $(L, \mathcal{D}(L))$. Then for $\phi \in \mathcal{D}(L) \cap C_b^+(E)$ (bounded positive continuous functions on E)

$$E_\mu[\exp(-\langle \phi, X_t \rangle)] = \exp(-\langle u_\phi(t), \mu \rangle)$$

where

$$(1)\qquad \begin{aligned} \frac{\partial u_\phi}{\partial t}(t) &= L u_\phi(t) - u_\phi(t)^2 \\ u_\phi(0) &= \phi \end{aligned}$$

It has long been known that when started from a finite initial measure the MVB process dies out (a.s.) and various schemes have been introduced to 'keep the process alive'. Several authors have studied the process conditioned on staying alive either up until some finite time or for ever (see e.g. Evans & Perkins [6], Roelly-Copolletta & Rouault [16], Etheridge [3]). Evans [5] produced a representation of the MVB process conditioned to stay alive for ever in terms of an 'immortal particle'. This particle moves around in space according to the spatial motion underlying the MVB process and throws off pieces of mass which go on to evolve as mass evolves for the unconditioned MVB process. Our purpose here is to produce a similar representation of the MVB process conditioned to stay alive until some finite time T. As $T \to \infty$, we recover the Evans representation. In the case when $L = \Delta$ (the Laplacian on R^n), our representation leads to a connection between the support of the conditioned superprocess at time T

and the profile at blow-up of solutions to the semilinear heat equation:

$$\frac{\partial u}{\partial t} = \Delta u + u^2$$

This equation has been extensively studied in the differential equations literature. The connection promises to offer some new insight.

We shall keep the description of our representation informal in the hope that it will be intuitively clear. A more formal approach would simply mimic Evans [5].

Let $\mathcal{N}_E$ denote the purely atomic measures on E. Our representation is as follows. Suppose that we wish to construct the MVB process from initial measure μ and conditioned to be alive at time T. We begin by introducing an $\mathcal{N}_E$-valued process $\{Y_t\}_{0\leq t<T}$. Y_t is a branching Markov process. Each atom of Y_t (independently of other atoms) has an exponential lifetime with time-variable parameter $V(t) = 1/(T-t), 0 \leq t < T$. Until they die, the atoms of Y_t evolve (independently) according to the spatial motion underlying the MVB process. When it dies, an atom is replaced (in the location where it died) by two offspring which go on to evolve (independently) in the same way as their parent. The atoms of Y_0 are located according to a Poisson random measure with intensity measure μ/T conditioned on there being at least one atom. We will refer to Y_t as the process of 'immortal Yule trees'. Having grown our immortal trees, we construct an $\mathcal{M}_F(E)$–valued process $\{Z_t\}_{0\leq t<T}$ as follows. $Z_0 = \mu$. Z_t evolves as the MVB process *conditioned to die before time* T but with immigration of mass at rate 2. The distribution of immigration at time t is Y_t. Then, for example, for $0 \leq s \leq t < T$, the expectation of the total amount of mass to be immigrated in the time interval $[s,t]$ is $2\int_s^t \langle 1, Y_r\rangle dr$. (Once immigrated mass evolves according to the 'killed' MVB process.) We will prove that $\{Z_t\}_{0\leq t<T}$ has the same distribution as the conditioned MVB process. [In the Evans representation, the distribution of immigration at time t is ξ_t where ξ_t is an atom moving according to the spatial motion underlying the MVB process. In this way the immortal Yule trees take the place of the Evans immortal particle.]

The representation is only valid for $t < T$ since we have said nothing about how to assign mass to the infinite number of immortals alive at time T. We claim the following:

Theorem 1. *Let $Z_t, t < T$ be the process described above, then*

$$Z_t \stackrel{d}{=} \{X_t | \langle 1, X_T\rangle \neq 0\}$$

i.e. for all $\phi \in \mathcal{D}(L) \cap C_b^+(E)$ and $t < T$

$$E_\mu[\exp(-\langle \phi, Z_t\rangle)] = E_\mu[\exp(-\langle \phi, X_t\rangle) | \langle 1, X_T\rangle \neq 0] \tag{2}$$

The proof of this result will just involve checking equation (2) and invoking uniqueness, so let us try to explain why it should be true. We begin by stating a lemma.

Lemma 2. *Fix $T > 0$. Given $X_0 \in \mathcal{M}_F(E)$, let $\xi_t^T = \xi_{t\wedge T}$ be an atom moving according to a copy of the spatial motion underlying the MVB process until stopped at time T. ξ_0^T is chosen uniformly according to X_0. Let W_t denote the MVB process conditioned to die before time T, but with immigration of mass at rate two with distribution ξ_t^T at time t. (Immigrated mass also evolves like the MVB process conditioned to die before time T.) Then*

$$W_t \stackrel{d}{=} \{X_t | \langle 1, X_T\rangle = 0, \langle 1, X_s\rangle \neq 0, 0 \leq s < T\}$$

Proof. The proof is a straightforward modification of Section 2 of Evans [5] and is omitted. (We immigrate at rate *two*, not rate one because Evans considers the MVB process with evolution equation $(\partial u)/(\partial t) = Lu - (1/2)u^2$.) ∎

One way to think of the branching structure of MVB with initial value μ is as a binary tree constructed under a Poisson number of (one-dimensional) Brownian excursions above zero. The intensity measure, n, of this Poisson process is normalised in such a way that $n(\{excursions\ with\ maximum\ height > \epsilon\}) = \langle 1, \mu\rangle/\epsilon$ (see Le Gall [9]). The excursion local time at level t then gives the mass of MVB at *time* t and, loosely, if two pieces of mass at time t correspond to times τ_1, τ_2 ($\tau_1 \leq \tau_2$) when a Brownian excursion B_τ is at level t, then they are conditionally independent given their position at time t_0 given by

$$t_0 = \inf\{B_\tau : \tau_1 \leq \tau \leq \tau_2\} \tag{3}$$

(**N.B.** The space variable for the Brownian excursion becomes the time variable for the MVB.)

Let us consider first a single Brownian excursion with maximum height S. This corresponds to a piece of measure-valued branching which dies at time S and from Lemma 2 we can think of this single upcrossing from 0 to S in terms of an immortal particle throwing off mass which goes on to perform the MVB process conditioned to die before time S. (In fact before becoming aware of Evans's work, the author produced a proof of Lemma 2 based on the discrete approximation to MVB obtained by approximating the Brownian excursions in the Le Gall picture by excursions of a simple random walk. c.f. Le Gall [10].)

Conditioning the process to have non-zero mass at time T is equivalent to conditioning on there being at least one excursion of the reflecting Brownian motion of height greater than T. Now consider one such excursion.

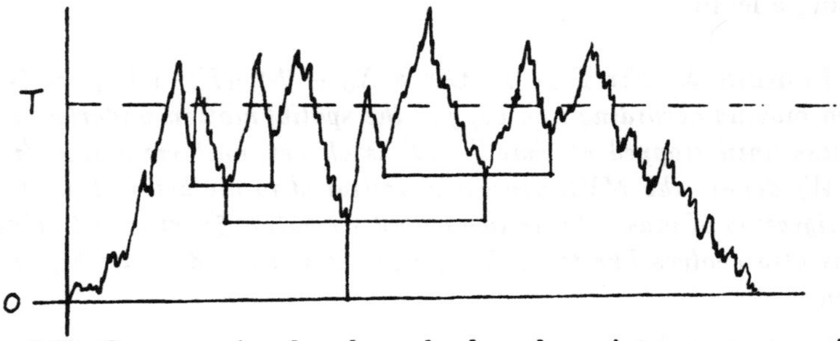

At level T, the excursion has been broken down into one upcrossing $0 \to T$, one downcrossing $T \to 0$, and a number of excursions below level T which don't hit zero. In the MVB setting, each such excursion is represented by an immortal particle, born at the infimum of the excursion, and throwing off mass as in Lemma 2. The immortal particles are not independent, but using (3) we can see how they are related. As indicated in the picture, they form a tree, and it is a simple consequence of the Williams path decomposition that this tree is a uniform 'stickbreaking process' (Le Gall [11]). Equivalently, the tree is a Yule process with branching rate $1/(T-t)$ at time t.

Since the number of excursions of height greater than T is Poisson with parameter $\langle 1, \mu\rangle/T$, this suggests our representation of the conditioned superprocess. Evidently there are a wealth of representations along these lines and in Etheridge [4] we will use this type of construction to produce measure-valued 'bridges'.

In Section 2 we prove our result and in Section 3 we outline the connection with the semilinear heat equation.

2. Proof of the representation

Our proof will consist in checking that the distribution of the process Z_t is precisely that of the MVB process conditioned on non-extinction at time T, so let us begin by recording this distribution. We will also require the distribution of the MVB process conditioned to die before time T.

Lemma 3. *For $\phi \in \mathcal{D}(L) \cap C_b^+(E)$ and $0 \leq t < T$,*

$$(a) \quad \begin{aligned} &E_\mu[\exp(-\langle \phi, X_t\rangle) | \langle 1, X_T\rangle \neq 0] \\ &\quad = \frac{(\exp(-\langle u_\phi(t), \mu\rangle) - \exp(-\langle u_{\phi+1/(T-t)}(t), \mu\rangle)}{(1 - \exp(-\langle 1/T, \mu\rangle))} \end{aligned}$$

$$(b) \quad E_\mu[\exp(-\langle \phi, X_t\rangle) | \langle 1, X_T\rangle = 0] = \exp(-\langle u_{\phi+1/(T-t)}(t) - 1/T, \mu\rangle)$$

where u_ψ solves the evolution equation (1) with $u_\psi(0) = \psi$.

Proof. The proof is trivial since for any finite measure ν and any time $S > 0$,

$$P_\nu[\langle 1, X_S\rangle = 0] = \lim_{\theta\to\infty} E_\nu[\exp(-\langle\theta, X_S\rangle)] = \exp(-\langle 1/S, \nu\rangle)$$

Thus for bounded measurable $F : \mathcal{M}_F(E) \to R$,

$$E_\mu[F(X_t)|\langle 1, X_T\rangle \neq 0] = \frac{E_\mu[F(X_t)(1 - \exp(-\langle 1/(T-t), X_t\rangle))]}{(1 - \exp(-\langle 1/T, \mu\rangle))}$$

and

$$E_\mu[F(X_t)|\langle 1, X_T\rangle = 0] = \frac{E_\mu[F(X_t)\exp(-\langle 1/(T-t), X_t\rangle)]}{\exp(-\langle 1/T, \mu\rangle)}$$

and the result follows. ∎

We work out the distribution of Z_t in stages. First of all, it is well known that the distribution of the branching Markov process Y_t (which forms our immortal trees) is determined by

$$E_\nu[\exp(\langle \log f, Y_t\rangle)] = \exp(\langle \log v(t), \nu\rangle)$$

where

$$\frac{\partial v}{\partial t} = Lv + \frac{1}{T-t}(v^2 - v) \qquad (4)$$
$$v(0) = f$$

(whenever a bounded positive solution exists. See, for example, Ikeda et al. [13].)

We will need an expression for the 'weighted occupation time' for Y.

Lemma 4. *For $0 \le t < T$ and $\psi \in \mathcal{D}(L) \cap C_b^+(E)$,*

$$E[\exp(-\int_0^t \langle\psi_s, Y_s\rangle ds)] = \exp(\langle \log w_t(t), Y_0\rangle)$$

where for $s \le t$

$$\frac{\partial w_t(s)}{\partial s} = Lw_t(s) + \frac{1}{(T-(t-s))}\Big(w_t(s)^2 - w_t(s)\Big) - \psi_{(t-s)}w_t(s) \qquad (5)$$
$$w_t(0) = 1$$

Proof. The proof is easily obtained by mimicking Iscoe [14] and is left to the reader. The only novelty is that since we take logarithms the equation for w_t does not take quite the form of the usual evolution equation for the weighted occupation time. In the usual set-up we would expect a '+' not a '−' in the last term. The '−' here arises from the decision to find an expression of the form $\exp(\langle \log w_t(t), Y_0\rangle)$ rather than $\exp(\langle -\log \tilde{w}_t(t), Y_0\rangle)$. By considering the integrated form of (5), one can convince oneself in the usual way that this equation has a unique solution with $0 \leq w_t \leq 1$. ∎

We are now in a position to write down the distribution of Z_t.

Proposition 5.

$$E[\exp(-\langle \phi, Z_t\rangle)] = \exp(-\langle u_{\phi+1/(T-t)}(t), \mu\rangle)\frac{(\exp(-\langle w_t(t)/T, \mu\rangle) - 1)}{(1-\exp(-\langle 1/T, \mu\rangle))}$$

where $w_t(t)$ is determined by

$$\begin{aligned} \frac{\partial w_t}{\partial s}(s) = L w_t(s) + \frac{1}{(T-(t-s))}\Big(w_t(s)^2 + w_t(s)\Big) \\ - 2u_{\phi+1/(T-t)}(s) w_t(s), \quad s \leq t \\ w_t(0) = 1 \end{aligned} \tag{6}$$

Proof. Recall first that if π_σ is a Poisson random measure with intensity measure σ, then

$$\int \exp(\langle f, \nu\rangle)\pi_\sigma(d\nu) = \exp(\langle e^f - 1, \sigma\rangle)$$

Moreover, the probability that there is at least one particle is $1 - \exp(-\langle 1, \sigma\rangle)$. By conditioning on $Y_s, 0 \leq s \leq t$, we see that in view of Lemma 3, the effect of the immigration (at rate two) is determined by Lemma 4 with

$$\psi_s = 2\Big(u_{\phi+1/(T-t)}(t-s) - \frac{1}{(T-s)}\Big)$$

Equation (5) then becomes equation (6) and using independence of the evolution of the initial mass, we have

$$\begin{aligned} E[\exp(-\langle \phi, Z_t\rangle)] &= \exp\big(-\langle u_{\phi+1/(T-t)} - (1/T), \mu\rangle\big) \\ &\quad \times \frac{(\exp(\langle w_t(t) - 1, \mu/T\rangle) - \exp(-\langle 1, \mu/T\rangle))}{1 - \exp(-\langle 1, \mu/T\rangle)} \end{aligned}$$

The result is now clear. ∎

By Lemma 3, the proof of the Theorem is now reduced to checking that

$$\exp(-\langle u_{\phi+1/(T-t)}(t),\mu\rangle)(\exp(-\langle w_t(t)/T,\mu\rangle)-1)$$
$$=\exp(-\langle u_\phi(t),\mu\rangle)-\exp(-\langle u_{\phi+1/(T-t)}(t),\mu\rangle)$$

or equivalently,

$$T\big(u_{\phi+1/(T-t)}-u_\phi\big)(t)=w_t(t)$$

Consider

$$V(s)=(T-(t-s))(u_{\phi+1/(T-t)}-u_\phi)(s)$$

It suffices to show that $V(t)=w(t)$. Now, using equation (1),

$$\frac{\partial}{\partial s}(u_{\phi+1/(T-t)}-u_\phi)(s)=L(u_{\phi+1/(T-t)}-u_\phi)(s)$$
$$+(u_{\phi+1/(T-t)}-u_\phi)(s)^2-2u_{\phi+1/(T-t)}(s)(u_{\phi+1/(T-t)}-u_\phi)(s)$$

Thus

$$\frac{\partial V}{\partial s}=LV+\frac{1}{(T-(t-s))}(V^2+V)-2u_{\phi+1/(T-t)}(s)V$$

But this is precisely the equation satisfied by $w_t(s)$ and $V(0)=1=w_t(0)$ and so by uniqueness of solutions to the differential equation, $V(t)=w_t(t)$ as required. ∎

Remark. If we let $T\to\infty$, then the branching rate of Y_t tends to zero (at any finite time), Y_t reduces to the Evans immortal particle and the effect of conditioning on extinction before time T disappears. i.e. We recover the Evans representation.

3. A semilinear heat equation

As was pointed out by Watanabe [17], if we take $E=R^n$ and $L=\Delta$ (the Laplacian), then for $\psi\in C_b^+(R^n)\cap\mathcal{D}(\Delta)$,

$$E[\exp(\langle\psi,X_t\rangle)]=\exp(\langle u(t),X_0\rangle)$$

where

$$\frac{\partial u}{\partial t} = \Delta u + u^2 \tag{7}$$
$$u(0) = \psi$$

and for $n < 2$, Fujita [8] proved that whenever $\{\psi(x) > 0\}$ has nonempty interior, the solution to this equation 'blows up' in a finite time, whereas if $n > 2$ (provided that ψ is bounded above by a small Gaussian) there are global solutions. Weissler [18] (amongst others) has since shown that $n = 2$ also belongs to the blow-up case. This equation has been extensively studied in the differential equations literature, with much of the work devoted to trying to understand the nature of the solution at blow-up.

Let T_0 denote the supremum of times for which the solution exists. A point x_0 is called a blow-up point if there exists a sequence $(x_n, t_n) \in R^n \times R_+$ such that

$$t_n \uparrow T_0, \quad x_n \to x_0, \quad u(x_n, t_n) \to \infty \qquad as \quad n \to \infty$$

If the initial condition ψ is sufficiently nice, then the rate of blow-up at such a point is known:

$$\lim_{t \uparrow T_0} (T_0 - t) u(t, x_0) = C \tag{8}$$

where C is a constant. (Under rather restrictive conditions on ψ, $C = 1$.) If x_0 is not a blow-up point, then the limit (8) is 0. This and related results can be found in Weissler [19],[20], Friedman & Mcleod [7], Baras & Cohen [1] or, more generally, Giga & Kohn [12]. Levine [15] reviews some of the known results for this (and related) equations and includes a substantial bibliography. Much current interest centres on establishing the 'profile' of u at the time of blow-up.

Our representation of the conditioned superprocess offers another probabilistic interpretation of equation (7). (Throughout this section $L = \Delta$.) The support of the superprocess at time T, is precisely the support of the (conditioned) Poisson number of branching Markov processes Y. This support is characterised as follows. Suppose that $f \in C(E)$ satisfies $0 \leq f \leq 1$. Then for $0 \leq t \leq T$,

$$\begin{aligned} &E_{\pi_{\mu/T}}[\exp(\langle \log f, Y_t \rangle) | \langle 1, Y_0 \rangle \neq 0] \\ &\qquad = \frac{(\exp(\langle -1 + v(t), \mu/T \rangle) - \exp(\langle -1, \mu/T \rangle))}{(1 - \exp(-\langle 1, \mu/T \rangle))} \\ &\qquad = \frac{\alpha_{\mu,T}}{(1 - \alpha_{\mu,T})} (\exp(\langle v(t), \mu/T \rangle) - 1) \end{aligned} \tag{9}$$

where $\alpha_{\mu,T} = \exp(-\langle 1, \mu/T\rangle)$, $v(0) = f$, $v(t)$ solves equation (4) for $t < T$ and $v(T) = \lim_{t\uparrow T} v(t)$.

Now for $t < T$, $v(t)$ may be written as

$$v(t) = (T - t)u(t)$$

where $u(t)$ solves (7) with $u(0) = f/T$. That $T \leq T_0$ where T_0 is the blow-up time for equation (7) for this initial condition (if one exists) is immediate by comparison with the initial condition $u(0) = 1/T$. Then we have

$$v(T) = \lim_{t\uparrow T}(T - t)u(t)$$

If $T = T_0$, then $v(T)$ in some sense describes the 'profile' of $u(\cdot)$ at blow-up.

In contrast to much of the analytic work, we need make no requirement on the decay of the initial condition or solution at infinity. In fact (9) remains valid for any f for which the right hand side converges. Thus for any initial condition ψ of equation (7) for which blow-up occurs at rate $(T_0 - t)^{-1}$ we can obtain this type of representation. Loosely, blow-up corresponds to there being 'too much mass in regions where ψ is large'.

Global existence of solutions to (7) for a given ψ corresponds to

$$E_{\pi_{\mu/T}}[\exp(\langle \log(T\psi), Y_T\rangle | \langle 1, Y_0\rangle \neq 0] = 0 \quad \forall T$$

Evidently, for this to be true, ψ will have to decay rather rapidly–which explains why existence of global solutions is only known for ψ less than a small Gaussian. Moreover, since most of the branches of Y are born near time T, and for large T there is exactly one immortal tree Y with high probability, one suspects (rightly) that if $\{\psi > 0\}$ has non-empty interior, then the underlying Brownian motion should be transient– i.e. $n \geq 3$.

As an example, suppose that we set $f = \chi_D$ (the indicator function of the (closed) set $D \subseteq E$). Then the left hand side of (9) reduces to

$$P_{\pi_{\mu/T}}[suppY_t \subseteq D | \langle 1, Y_0\rangle \neq 0]$$

and as $t \uparrow T$, this converges to

$$P_\mu[suppX_T \subseteq D | \langle 1, X_T\rangle \neq 0]$$

(A formal proof of this uses the fact that if the spatial motion underlying the MVB process is Brownian motion then the support of the MVB process is well-behaved.)

In this case we must consider the 'weak' (integrated) form of equations (4)and (7)

$$v(t) = T_t v(0) + \int_0^t \frac{1}{(T-s)} T_{t-s}\left(v(s)^2 - v(s)\right) ds$$

$$u(t) = T_t u(0) + \int_0^t T_{t-s}\big(u(s)^2\big)ds$$

where T_t denotes the semigroup generated by Δ. The probability that the conditioned superprocess charges an open set A at a finite time, is obtained by setting $D = A^c$ (to calculate one minus the probability). Now for any unbounded set D, the blow-up time of the solution to (7) with initial condition $u(0) = \chi_D$ is one. (To see this, transform the equation using the mapping $x \mapsto x^{-1}$, and one sees that blow-up for the transformed equation occurs at the origin at time one. As already remarked, comparison with the initial condition u(0)=1, shows that no blow-up occurs before time one.) From this picture, one sees that for initial conditions of the form $u(0) = \chi_D$ where D is unbounded, the blow-up set will in general be the whole of R^d, and in contrast to the solution for initial conditions more amenable to analytic attack, the 'profile' of u at blow-up will be non-trivial.

Acknowledgement: The author is currently Neyman Assistant Professor at the Department of Statistics, U.C. Berkeley. Thanks to everyone there for their (continuing) hospitality.

REFERENCES

[1] P.Baras & L.Cohen (1987). Complete blow-up after T_{max} for the solution of a semilinear heat equation. *J. Functional Anal.* **71** *142-174*

[2] D. A. Dawson (1991). Measure-valued Markov processes. **LNM** *To appear.*

[3] A. M. Etheridge (1989). Asymptotic behaviour of some measure-valued diffusions. *Oxford D. Phil Thesis.*

[4] A. M. Etheridge (1992). Measure-valued branching bridges. *In preparation.*

[5] S. N. Evans (1992). Two representations of a conditioned superprocess. *Preprint.*

[6] S. N. Evans & E. A. Perkins (1990). Measure-valued Markov branching processes conditioned on non-extinction. *Israel J. Math* **71** *329-337*

[7] A.Friedman & B.McLeod (1985). Blow-up of positive solutions of semilinear heat equations. *Indiana Univ. Math. J.* **34** *425-447*

[8] H.Fujita (1966). On the blowing up of solutions of the Cauchy problem for $u_t = \Delta u + u^{1+\alpha}$. *J. Fac. Sci. Univ. Tokyo Sect. A Math* **16** *105-113*

[9] J. F. Le Gall (1991). Brownian excursions, trees and measure-valued branching processes. *Ann. Prob.* **19** *1399-1439*

[10] J. F.Le Gall (1986). Une approche elementaire des théorèmes de décomposition de Williams. *Sem. Prob. XX* **LNM 1204** *Springer-Verlag*

[11] J. F.Le Gall (1989). Marches aléatoires, mouvement brownien et processus de branchement. *Sem. Prob. XXIII* **LNM 1372** *Springer-Verlag*

[12] Y.Giga & R. V. Kohn (1987). Characterizing blowup using similarity variables. *Indiana Univ. Math. J.* **36** *1-40*

[13] N. Ikeda, M. Nagasawa & S. Watanabe (1968), (1969). Branching Markov processes I,II,III. *J. Math. Kyoto Univ.* **8** *233-278* **9** *95-160*

[14] I.Iscoe (1986). A weighted occupation time for a class of measure-valued branching processes. *Prob. Th. Rel. Fields* **71** *85-116*

[15] H. A. Levine (1990). The role of critical exponents in blow-up theorems. *SIAM Review* **32 no. 2** *262-288*

[16] S. Roelly-Coppoletta & A.Rouault (1989). Processus de Dawson-Watanabe conditionné par le futur lointain. *C. R. Acad. Sci. Paris* **t. 309** *Série I 867-872*

[17] S.Watanabe (1968). A limit theorem of branching processes and continuous state branching. *J. Math. Kyoto Univ.* **8** *141-167*

[18] F. B.Weissler (1981). Existence and non-exixtence of global solutions for a semilinear heat equation. *Israel J. Math* **38** *29-40*

[19] F. B.Weissler (1984). Single point blow-up for a semilinear initial value problem. *J. Diff. Eq.* **55** *204-224*

[20] F. B.Weissler (1985). An L^∞ blow-up estimate for a nonlinear heat equation. *Comm. Pure Appl. Math* **38** *292-295*

Department of Mathematics and Statistics
University of Edinburgh
U.K.

Markovian Bridges: Construction, Palm Interpretation, and Splicing

by

Pat Fitzsimmons[1], Jim Pitman[2], Marc Yor[2]

1. Introduction.

By a Markovian bridge we mean a process obtained by conditioning a Markov process X to start in some state x at time 0 and arrive at some state z at time t. Once the definition is made precise, we call this process the (x,t,z)-bridge derived from X. Important examples are provided by Brownian and Bessel bridges, which have been extensively studied and find numerous applications. See for example [**PY1, SW, Sa, H, EL, AP, BP**]. It is part of Markovian folklore that the right way to define bridges in some generality is by a suitable Doob h-transform of the space-time process. This method was used by Getoor and Sharpe [**GS4**] for excursion bridges, and by Salminen [**Sa**] for one-dimensional diffusions, but the idea

[1] Research supported in part by NSF Grant DMS 91-01675.
[2] Research supported in part by NSF Grant DMS 91-07531.

of using h-transforms to construct bridges seems to be much older. Our first object in this paper is to make this definition of bridges precise in a suitable degree of generality, with the aim of dispelling all doubts about the existence of clearly defined bridges for nice Markov processes. This we undertake in Section 2. In Section 3 we establish a conditioning formula involving bridges and continuous additive functionals of the Markov process. This formula can be found in [**RY**, Ex. (1.16) of Ch. X, p.378] under rather stringent continuity conditions. One of our goals here is to prove the formula in its "natural" setting. We apply the conditioning formula in Section 4 to show how Markovian bridges are involved in a family of Palm distributions associated with continuous additive functionals of the Markov process. This generalizes an approach to bridges suggested in a particular case by Kallenberg [**K1**], and connects this approach to the more conventional definition of bridges adopted here.

Our concern with these matters was prompted by an interesting splicing construction, the inverse of a path decomposition that we now describe in the context where we first encountered it. Perman, Pitman and Yor [**PPY**] considered a process $Y = (Y_s)_{0\le s\le 1}$ constructed as follows from a standard Brownian bridge $B = (B_s)_{0\le s\le 1}$, that is to say the $(0,1,0)$-

bridge derived from a standard one-dimensional Brownian motion. Let U be a random time independent of B, with uniform distribution on $[0,1]$. Let G be the time of the last zero of B before time U, D the time of the first zero of B after time U. Delete the segment of the path of B over the excursion interval $]G,D[$, close up the gap of length $D-G$ to obtain a path of random length $1-(D-G)$, then standardize this path by Brownian scaling to obtain a random path Y parameterized by $[0,1]$. According to [**PPY**, Corollary 3.15], the law of this process Y is absolutely continuous with respect to the law of the standard Brownian bridge B, with density proportional to L_1, the local time of B at zero up to time 1. On the other hand, Aldous and Pitman [**AP**] give a decomposition of the original Brownian bridge B into three independent paths: two standard bridges obtained by Brownian scaling of the segments of B on $[0,G]$ and B on $[D,1]$, and one standard signed excursion obtained similarly from B on $[G,D]$; moreover these three paths of length 1 are independent of the triple of interval lengths $(G, D-G, 1-D)$, which has the same exchangeable Dirichlet(1/2,1/2,1/2) distribution as $(N^2/\Sigma, M^2/\Sigma, O^2/\Sigma)$, where N, M and O are independent standard normal variables, and $\Sigma = N^2 + M^2 + O^2$. Since the path of Y can be recovered from the two bridges in this decomposition and the fraction $T = \frac{G}{G+1-D}$, it follows that Y admits the following decomposition at time T:

(i) $\mathbb{P}(T \in dt) = \rho(t)dt$, where $\rho(t) = (\pi\sqrt{t(1-t)})^{-1}$ is the arc-sine density on $[0,1]$ of $N^2/(N^2+O^2)$, and

(ii) conditional on the event $\{T = t\}$, Y splits into independent pieces $(Y_s)_{0\leq s\leq t}$ and $(Y_s)_{t\leq s\leq 1}$ whose conditional distributions are those of Brownian bridges over the time intervals $[0,t]$ and $[t,1]$ respectively.

Here by a Brownian bridge over $[u,v]$, say, we mean a standard Brownian motion started at 0 at time u and conditioned to return to 0 at time v.

The above observation prompts the question of what process Y is obtained by prescribing some other density $\rho(t)$ on $[0,1]$ for T as in (i) above, then splicing together two Brownian bridges of lengths T and $1-T$, as in (ii)? The law of Y can be computed explicitly:

$$\text{(1.0)} \qquad \mathbb{P}(Y \in d\omega) = \int_0^1 f(t)\, dL_t(\omega) \cdot \mathbb{P}(B \in d\omega),$$

where B denotes a Brownian bridge, $L_t(\omega)$ is the Brownian local time at zero up to time t, defined for all $0 \leq t \leq 1$ for almost all ω relative to the law of B, and $f(t) = \rho(t)\sqrt{2\pi t(1-t)}$. Using a version of Bayes' rule one can compute the conditional distribution of T given Y:

$$(1.1) \qquad \mathbb{P}(T \in dt \mid Y) = \frac{f(t)\, dL_t(Y)}{\int_0^1 f(s)\, dL_s(Y)}.$$

In particular, in case $\rho(t)$ is the arcsine density, $f(t) = \sqrt{2/\pi}$ is constant, the density of the law of Y relative to B reduces to $\sqrt{2/\pi}L_1$, and the conditional distribution of T given Y is proportional to the local time: $P(T \leq t|Y) = L_t(Y)/L_1(Y)$. Note also that no such splicing of bridges can yield a "pure" bridge Y, for the density factor in (1.0) cannot be 1 almost surely, no matter what the choice of $\rho(t)$. Put another way, given a standard Brownian bridge B, it is impossible to find a random time T such that T falls almost surely in the zero set of B, and given T the processes B on $[0,T]$ and B on $[T,1]$ are independent bridges over these intervals.

In the main result of this paper, we apply the Palm interpretation of bridge distributions to establish a general "splicing" construction for Markov processes which includes the above example as a special case. An informal description of this result is as follows. Let $(X_s)_{s\geq 0}$ be a time-homogeneous Markov process with state space E. Fix $\ell > 0$ and let a space-time point (Z,T) be chosen according to a suitable distribution on $E \times [0,\ell]$. Conditional on $\{Z = z, T = t\}$ let Y be the concatenation of independent X-bridges, the first an

(x,t,z)-bridge, the second a $(z,\ell-t,y)$-bridge. Then the (unconditional) distribution of Y is absolutely continuous with respect to the law of an (x,ℓ,y)-bridge, and the Radon-Nikodym derivative can be written down explicitly. Moreover, the conditional distribution of (Z,T) given Y can be expressed by a formula analogous to (1.1). This result is presented more formally as Proposition 4 in Section 4. Finally, in Section 5 we record a general probabilistic interpretation of the family of Palm measures associated with a random measure, which underlies the splicing result in the context of Markovian bridges.

2. Construction of Bridges

For the rest of the paper, we will work with a right Markov process $X = (X_t)_{t\geq 0}$ with state space E and transition semigroup (P_t). Thus X is a strong Markov process with right continuous sample paths. To streamline the exposition, we assume that E is Lusinian (i.e., homeomorphic to a Borel subspace of some compact metric space), that P_t maps Borel functions to Borel functions, and that the paths of X are cadlag. This allows us to realize X as the coordinate process on the sample space Ω of all cadlag paths from $[0,\infty[$ to E. The law of X when started at x is $\mathbb{P}_x$. We write $(\mathcal{F}_t)_{t\geq 0}$ for the natural (uncompleted) filtration of X and $(\theta_t)_{t\geq 0}$ for the usual shift operators. (Certain details, in particular the completion of the σ-algebras $\mathcal{F}_t$, will be left to the reader.)

A reasonable theory of bridges requires (at a minimum) that X have transition densities; namely that

$$P_t(x, dy) = p_t(x, y)\, m(dy) \tag{2.1}$$

where m is a σ-finite measure on E. Note that this condition forbids jumps at fixed times:

$$\mathbb{P}_x(X_t = X_{t-}) = 1, \quad \forall t > 0, x \in E. \tag{2.1a}$$

To see this, consider T, the first time at which X has a jump of size bigger than ϵ; then by virtue of (2.1), $\mathbb{P}_x(T = t) = 0$ for all $t > 0$ by [**GS4**, (3.18)].

To develop things fully we need to assume a bit more. We suppose that there is a second right process $\hat{X}$ in duality with X relative to the measure m. This means that the semigroup $(\hat{P}_t)$ of $\hat{X}$ is related to (P_t) by

$$\int_E f(x) P_t g(x)\, m(dx) = \int_E \hat{P}_t f(x) g(x)\, m(dx), \tag{2.2}$$

for all $t > 0$ and all positive Borel functions f and g. For simplicity we assume that X and $\hat{X}$ have infinite lifetimes; this implies that m is an invariant measure for X (and for $\hat{X}$ as well). It is known that (2.1) and (2.2) imply that there is a version of the density $p_t(x, y)$ that is jointly measurable in (t, x, y) and such that the Chapman-Kolmogorov identity

$$p_{t+s}(x, y) = \int_E p_t(x, z) p_s(z, y)\, m(dz) \tag{2.3}$$

holds for all $s, t > 0$, and $x, y \in E$. Moreover, the dual of (2.1) is valid:

$$\hat{P}_t(x, dy) = p_t(y, x)\, m(dy). \tag{2.4}$$

See [**D, GS4, Wi, Y**].

Note, for example, that any one-dimensional regular diffusion without absorbing boundaries satisfies the above hypotheses, with the speed measure of the diffusion serving as the reference measure m ; see [**IM**, p. 149 ff.].

We now use Doob's method of h-transforms to construct bridge laws $\mathbb{P}^{\ell}_{x,y}$, which for each x and ℓ will serve as a family of regular $\mathbb{P}_x$ conditional laws for $(X_t, 0 \le t < \ell)$ given $X_{\ell-} = y$. In view of (2.1a), these $\mathbb{P}^{\ell}_{x,y}$ will serve equally well as conditional laws given $X_\ell = y$ rather than given $X_{\ell-} = y$. But with the h-transform approach it is natural to think primarily in terms of conditioning the left limit. For background and further details, the reader can consult [**M1, F, AJ, GS1, GS4, RY**].

Fix $x, y \in E$ and $\ell > 0$ such that $0 < p_\ell(x,y) < \infty$. Using (2.3) it is a simple matter to check that the process

$$H_t = p_{\ell-t}(X_t, y), \quad 0 \le t < \ell,$$

is a (positive) martingale under $\mathbb{P}_x$. Consequently, the formula

$$(2.5) \qquad Q(A) = \int_A H_t(\omega)\,\mathbb{P}_x(d\omega), \qquad A \in \mathcal{F}_t, \quad 0 \le t < \ell,$$

defines a finitely additive set function $Q = Q^\ell_{x,y}$ on the algebra $\mathcal{G} = \cup_{0 \le t < \ell} \mathcal{F}_t$ such that each restriction $Q|_{\mathcal{F}_t}$ is σ-additive. We claim that Q extends to a measure on $\mathcal{F}_{\ell-}$, the σ-algebra generated by $\mathcal{G}$. This extension, when normalized by $p_\ell(x,y)$, will be the law $\mathbb{P}^\ell_{x,y}$.

We verify the claim as follows. Let Ω_ℓ be the space of right continuous paths from $[0,\ell[$ to E that have left limits on $]0,\ell[$. We can view Q as a finitely additive measure on Ω_ℓ in the obvious way. The point is that Ω_ℓ equipped with its natural filtration $(\mathcal{G}_t)_{0 \le t < \ell}$ is "projectively closed" so that we may apply the projective limit theorem to conclude that Q extends to a (σ-additive) measure Q^* on the σ-algebra $\vee_{0 \le t < \ell} \mathcal{G}_t$. (See **[AJ]** or the appendix in **[F]**; $(\Omega_\ell, (\mathcal{G}_t)_{0 \le t < \ell})$ is a "standard system" in the terminology of Parthasarathy **[Pa]**.) If we knew that Q^* gave full measure to the set of paths with left limits at ℓ, then we could identify Q^* with a measure on $(\Omega, \mathcal{F}_{\ell-})$ and this would be the desired extension. Here is the first place where the duality hypothesis (2.2) comes into play. For we can make the dual construction, obtaining a measure $\hat{Q}^*$ on Ω_ℓ corresponding to $\hat{X}$ started at y and conditioned to have left limit x at time ℓ. Let Ω_ℓ^+ be the space of cadlag maps from $]0,\ell[$ to E. We can view both Q^* and $\hat{Q}^*$ as measures on Ω_ℓ^+. Using (2.2), a check of finite dimensional distributions shows that Q^* is the image of $\hat{Q}^*$ under the time-reversal mapping which sends a path $\omega \in \Omega_\ell$ to the path $(\omega((\ell - t)-))_{0<t<\ell}$. Since $\hat{Q}^*$ is concentrated on those paths whose right limits at time 0 exist and equal y, it must be that Q^* concentrates its mass on those paths whose left limits at time ℓ exist and equal y.

Thus there is a measure $Q = Q^\ell_{x,y}$ on $(\Omega, \mathcal{F}_{\ell-})$ such that (2.5) holds. Let us define

$$\mathbb{P}^\ell_{x,y} = \begin{cases} [p_\ell(x,y)]^{-1}\, Q^\ell_{x,y}, & \text{if } 0 < p_\ell(x,y) < \infty, \\ 0, & \text{otherwise.} \end{cases}$$

The above discussion is summarized in the following

Proposition 1. *Under the assumptions (2.1), (2.2), and (2.3), if $0 < p_\ell(x,y) < \infty$ then there is a unique probability measure $\mathbb{P}^\ell_{x,y}$ on $(\Omega, \mathcal{F}_{\ell-})$ such that*

$$\mathbb{P}^\ell_{x,y}(F) \cdot p_\ell(x,y) = \mathbb{P}_x(F \cdot p_{\ell-t}(X_t, y)) \tag{2.6}$$

for all positive $\mathcal{F}_t$*-measurable functions* F *on* Ω*, for all* $0 \leq t < \ell$*. Under* $\mathbb{P}^{\ell}_{x,y}$*, the coordinate process* $(X_t)_{0 \leq t < \ell}$ *is a non-homogeneous strong Markov process with transition densities*

(2.7)
$$p^{(y,\ell)}(z, s; z', t) = \frac{p_{t-s}(z, z')p_{\ell-t}(z', y)}{p_{\ell-s}(z, y)}, \qquad 0 < s < t < \ell.$$

Moreover $\mathbb{P}^{\ell}_{x,y}(X_0 = x, X_{\ell-} = y) = 1$*. Finally, if* $F \geq 0$ *is* $\mathcal{F}_{\ell-}$*-measurable, and* $g \geq 0$ *is a Borel function on* E*, then*

$$\text{(2.8)} \qquad \mathbb{P}_x(F\, g(X_{\ell-})) = \int_E \mathbb{P}^{\ell}_{x,y}(F)\, g(y)\, p_\ell(x,y)\, m(dy),$$

a formula which holds also with X_ℓ *instead of* $X_{\ell-}$*. Thus* $(\mathbb{P}^{\ell}_{x,y})_{y \in E}$ *is a regular version of the family of conditional probability distributions* $\mathbb{P}_x(\cdot \,|X_{\ell-} = y)$*,* $y \in E$*, equally so with* X_ℓ *instead of* $X_{\ell-}$*.*

Proof. The strong Markov property of $(X_t)_{0 \leq t < \ell}$ under $\mathbb{P}^{\ell}_{x,y}$ follows from (2.6) by optional stopping, and the reader can easily verify the formula (2.7) for the transition densities. Thanks to the monotone class theorem, in proving (2.8) it suffices to consider $\mathcal{F}_t$-measurable F, where $0 < t < \ell$. For such F, by the Markov property at time t, (2.6), and (2.1a),

$$
\begin{aligned}
\mathbb{P}_x(Fg(X_{\ell-})) &= \mathbb{P}_x(F\,P_{\ell-t}\,g(X_t)) \\
&= \mathbb{P}_x(F\int p_{\ell-t}(X_t,y)g(y)\,m(dy)) \\
&= \int \mathbb{P}_x(F\,p_{\ell-t}(X_t,y))\,g(y)\,m(dy) \\
&= \int \mathbb{P}^{\ell}_{x,y}(F)p_\ell(x,y)\,g(y)\,m(dy),
\end{aligned}
$$

as desired. □

Remark. A few comments concerning ratios such as (2.7) are in order. First note that if $p_\ell(x,y) < \infty$, then the optional sampling theorem applied to the martingale $H_t = p_{\ell-t}(X_t,y)$ shows that $\{(z,t) : p_{\ell-t}(z,y) = \infty\}$ is $\mathbb{P}_x$-polar in the sense that $\mathbb{P}_x(p_{\ell-t}(X_t,y) = \infty$ for some $t \in]0,\ell\,[) = 0$. Because of (2.6), it is likewise true that $\{(z,t) : p_{\ell-t}(z,y) = \infty\}$ is $\mathbb{P}^{\ell}_{x,y}$-polar. Similarly, since a positive martingale "sticks" to the value 0 once that value is attained, we have $\mathbb{P}^{\ell}_{x,y}(p_{\ell-t}(X_t,y) = 0$ for some $t \in [0,\ell\,[) = 0$. These observations should comfort the reader should our subsequent treatment of ratios such as (2.7) seem cavalier.

We conclude this section by mentioning some results which flow easily from Proposition 1 under the basic hypotheses (2.1), (2.2) and (2.3).

Corollary 1. *Suppose* $0 < p_\ell(x,y) < \infty$. *The* $\mathbb{P}^{\ell}_{x,y}$*-law of the time-reversed process* $(X_{(\ell-t)-})_{0\le t<\ell}$ *is* $\hat{\mathbb{P}}^{\ell}_{y,x}$, *the law of* a (y,ℓ,x)-*bridge for the dual process* $\hat{X}$.

Corollary 2. *Suppose* $0 < p_\ell(x,y) < \infty$. *For each* $(\mathcal{F}_t)$ *stopping time* T, *a* $\mathbb{P}^\ell_{x,y}$ *regular conditional distribution for* $(X_{T+u}, 0 \le u < \ell - T)$ *given* $\mathcal{F}_T$ *on* $(T < \ell)$ *is provided by* $\mathbb{P}^{\ell-T}_{X_T,y}$.

Applied to the dual process after time reversal and conditioning on X_ℓ, Corollary 2 implies the following decomposition of the original Markov process X at random times τ that correspond to stopping times on the reversed time scale (first for τ with $\tau < \ell$, then for unbounded τ by an easy argument). A special case of this result is due to Kallenberg [**K1**]. See also [**Yo**] for an application to Bessel processes.

Corollary 3. *Suppose that* X *is a Hunt process. For each random time* τ *such that* $(\tau \ge t)$ *is in the* σ*-field generated by* $(X_{t+u}, u \ge 0)$, *a* $\mathbb{P}_x$ *regular conditional distribution for* $(X_t, 0 \le t < \tau)$ *given* $(\tau, X_{\tau-}, X_{\tau+u}, u \ge 0)$, *is provided by* $\mathbb{P}^\tau_{x,X_{\tau-}}$.

A crucial ingredient of the proof is the "left Markov property" of X at such times τ, which are co-optional relative to the space-time process based on X. See [**GS2, GS3**] for details, especially section 6 of [**GS3**].

3. Conditioning Formula.

Recall that a continuous additive functional (CAF) of X is a continuous increasing adapted process $(A_t)_{t\geq 0}$ such that for all $s, t \geq 0$, for $\mathbb{P}_x$ a.e. $\omega \in \Omega$,

$$A_{t+s}(\omega) = A_t(\omega) + A_s(\theta_t \omega). \tag{3.1}$$

By a "perfection" theorem due to Walsh [**Wa**], we can assume without loss of generality that (3.1) holds for all $s, t \geq 0$ and all $\omega \in \Omega$, that $t \mapsto A_t(\omega)$ is continuous for all $\omega \in \Omega$, and that A_t is $\mathcal{F}^*_{t+}$-measurable for each $t \geq 0$, where $\mathcal{F}^*_t$ is the universal completion of $\mathcal{F}_t = \sigma\{X_s, 0 \leq s \leq t\}$. See also the appendix in [**G**] for a detailed discussion.

In the following discussion we fix a CAF $A = (A_t)$such that

$$\mathbb{P}_x(A_t < \infty) = 1, \quad \forall t > 0, x \in E. \tag{3.2}$$

All that follows is based on a formula, due to Revuz [**R**], allowing the explicit computation of expectations involving A. The *Revuz measure* ν associated with A is defined by

$$\nu(g) = t^{-1} \mathbb{P}_m \left(\int_0^t g(X_s)\, dA_s \right), \tag{3.3}$$

where g is any positive Borel function on E. The fact that the right side of (3.3) does not depend on $t > 0$ is an easy consequence of (3.1) and the invariance of m. Because of (3.2), the measure ν is σ-finite; see [**R**, p. 509]. The role of ν is expressed in the following key formula: if f is a positive measurable function on $E \times [0,\infty[$, then for all $x \in E$ and $t > 0$,

(3.4)

$$\mathbb{P}_x\left(\int_0^t f(X_s,s)\,dA_s\right) = \int_0^t ds \int_E \nu(dy)\,p_s(x,y)f(y,s).$$

Actually, the Laplace-transformed version of this formula (under the hypothesis of resolvent duality) was established by Revuz [**R**] for standard processes, and under weaker conditions by Meyer [**M2**]. Getoor & Sharpe [**GS4**] have proved formula (3.4) under the present hypotheses (plus standardness) by applying Revuz' result to the space-time process (X_t,t), which is in resolvent duality with $(\hat{X}_t,-t)$. Since we are restricting attention to continuous additive functionals, standardness is unnecessary. See [**AM**] and also [**GS1**].

Before proceeding we record the version of (3.4) appropriate to bridges.

Lemma 1. *Fix $\ell > 0$ and $x,y \in E$ such that $0 < p_\ell(x,y) < \infty$. Then for $0 \le t \le \ell$ and Borel $f \ge 0$,*

$$\mathbb{P}^\ell_{x,y}\left(\int_0^t f(X_s,s)\,dA_s\right)$$

$$(3.5) \qquad = \int_0^t ds \int_E \nu(dz) \left[\frac{p_s(x,z)p_{\ell-s}(z,y)}{p_\ell(x,y)} \right] f(z,s).$$

Proof. It suffices to prove the lemma in case $0 < t < \ell$. Using (2.6) we have

$$p_\ell(x,y)\, \mathbb{P}^\ell_{x,y} \left(\int_0^t f(X_s, s)\, dA_s \right)$$

$$(3.6) \qquad = \mathbb{P}_x \left(\int_0^t f(X_s,s)\, dA_s \cdot p_{\ell-t}(X_t, y) \right).$$

But for fixed t the $\mathbb{P}_x$-optional projection of the (constant in time) process $s \mapsto p_{\ell-t}(X_t, y)$, $0 \le s \le t$, is the process $s \mapsto P_{t-s}(X_s, p_{\ell-t}(\cdot, y))$, $0 \le s \le t$, which coincides with $p_{\ell-s}(X_s, y)$ because of the Chapman-Kolmogorov identity. Feeding this into (3.6), and then using (3.4), we arrive at the right side of (3.5), and the lemma is proved. □

We now prove our CAF conditioning formula, first for the laws $\mathbb{P}_x$.

Proposition 2. *If $H \ge 0$ is a predictable process and $f \ge 0$ is a Borel function on $E \times [0, \infty[$, then*

$$\mathbb{P}_x \left(\int_0^t H_s\, f(X_s, s)\, dA_s \right)$$

$$(3.7) \qquad = \mathbb{P}_x \left(\int_0^t \mathbb{P}^s_{x,X_s}(H_s)\, f(X_s,s)\, dA_s \right).$$

Proof. By the monotone class theorem we need only consider processes H of the form $H_s = 1_{]a,b]}(s) \cdot C$, where $0 \le a < b \le t$

and $C \in \mathcal{F}_a$ is bounded and positive. For such an H the left side of (3.7) becomes

$$\mathbb{P}_x\left(C\cdot\int_a^b f(X_s,s)\,dA_s\right)$$

$$= \mathbb{P}_x\left(C\cdot\left(\int_0^{b-a} f(X_u,u+a)\,dA_u\right)\circ\theta_a\right)$$

$$(3.8)\qquad = \mathbb{P}_x\left(C\cdot\mathbb{P}_{X_a}\left(\int_0^{b-a} f(X_u,u+a)\,dA_u\right)\right).$$

On the other hand the right side of (3.7) equals

(3.9)

$$\mathbb{P}_x\left(\int_a^b \mathbb{P}^s_{x,X_s}(C)\,f(X_s,s)\,dA_s\right)$$

$$= \int_a^b ds\int_E \nu(dz)\mathbb{P}^s_{x,z}(C)\,p_s(x,z)\,f(z,s), \quad \text{by (3.4)}$$

$$= \int_a^b ds\int_E \nu(dz)\mathbb{P}_x(C\,p_{s-a}(X_a,z))\,f(z,s), \quad \text{by (2.6)}$$

$$= \mathbb{P}_x(C\,g(X_a)),$$

where

$$g(y) = \int_a^b ds\int_E \nu(dz)p_{s-a}(y,z)f(z,s)$$

$$= \mathbb{P}_y\left(\int_0^{b-a} f(X_u,u+a)\,dA_u\right).$$

Thus (3.8) and (3.9) combine to yield (3.7). □

Remark. Since the diffuse measure dA_s does not charge the countable set of discontinuities of X, formula (3.7) holds also with X_s replaced by X_{s-} (three times). This point should be borne in mind in the sequel. It is also worth noting that Proposition 2 is valid if X is a Hunt process and A is a *predictable* (but possibly discontinuous) additive functional, provided X_s is replaced by X_{s-} in (3.3) and (3.7). (See [**GS4**].) In fact, the same replacement makes (3.4) valid for predictable A, and the reader will have noticed that the continuity of A was not used explicitly in the proof of Proposition 2. Any Hunt process satisfying (2.1) (but not necessarily the duality hypothesis (2.2)) for which bridge laws $\mathbb{P}^{\ell}_{x,y}$ can be constructed such that the altered form of (3.7) holds for all predictable H and A, must satisfy the CMF (=Co-Markovien Forte) condition of Azéma [**A**]. If, in addition, the reference measure m is invariant, then by results in [**A**], X must have a strong Markov dual $\hat{X}$ such that (2.2) holds. In this sense our duality hypothesis is not too much to ask if we wish (3.7) to hold.

The bridge form of (3.7) follows from Proposition 2 in the same way that Lemma 1 followed from (3.4). We state the result but leave the proof to the reader.

Proposition 3. *If $H \geq 0$ is a predictable process and $f \geq 0$ is a Borel function on $E \times [0, \infty[$, and if $0 < p_\ell(x, y) < \infty$, then for $0 \leq t \leq \ell$,*

$$\mathbb{P}^{\ell}_{x,y}\left(\int_0^t H_s\, f(X_s, s)\, dA_s\right)$$

$$(3.10) \qquad = \mathbb{P}^{\ell}_{x,y}\left(\int_0^t \mathbb{P}^{s}_{x,X_s}(H_s)\, f(X_s,s)\, dA_s\right).$$

More general versions of Propositions 2 and 3, in which $\int_0^t f(X_s,s)\,dA_s$ is replaced by an arbitrary predictable additive functional of the space-time process, are also valid if X is a Hunt process. The ramified form of Propostion 3 leads to an alternate proof of Corollary 3 once one notes that for a time τ as in the corollary, the dual predictable projection of the increasing process $t \mapsto 1_{\{\tau \le t\}}$ is an additive functional of the space-time process. (Recall from the remark following Corollary 3 that τ is co-optional for the space-time process.) In the same vein one can express the conditional distribution of $(X_t, 0 \le t < \tau)$ given $(X_{\tau-}, \tau)$ in terms of bridge laws, provided the dual predictable projection of $t \mapsto 1_{\{0<\tau\le t\}}$ has the form $\int_0^t \lambda_s\, dAs$, where λ is a predictable process and A is a predictable additive functional. See [**JY**, Prop. (3.13)] for a prototype result in this direction.

4. Splicing Bridges.

We are now prepared to prove the splicing theorem outlined in section 1. Fix $x, y \in E$ and $\ell > 0$ such that $0 < p_\ell(x,y) < \infty$. Also, fix a probability distribution on $E \times]0,\ell[$ of the form $\rho(z,t)\,\nu(dz)\,dt$, where $\rho(z,t) = 0$ whenever $p_t(x,z)p_{\ell-t}(z,y) = 0$, and where ν is the Revuz measure of a CAF $A = (A_t)$ as in section 3. (Revuz [**R**] has shown that a measure ν is associated with a CAF satisfying (3.2) if and only if ν charges no semipolar set and there is a strictly positive Borel function ψ on E such that

$$\sup_{x \in E} \int_0^\infty \exp(-t) p_t(x,y)\, \psi(y)\, \nu(dy) < \infty$$

and $\inf\{t > 0 : \psi(X_t) \le 1/n\} \to \infty$ as $n \to \infty$, almost surely.) Consider the probability space $(\Omega^*, \mathcal{F}^*, \mathbb{P}^*)$:

$$\Omega^* = E \times]0, \ell[\times \Omega \times \Omega,$$

$$\mathcal{F}^* = \mathcal{E} \otimes \mathcal{B}(]0, \ell[) \otimes \mathcal{F}_{\ell-} \otimes \mathcal{F}_{\ell-},$$

$$\mathbb{P}^*(dz, dt, d\omega, d\omega') = \rho(z,t)\, \nu(dz)\, dt\, \mathbb{P}^t_{x,z}(d\omega) \mathbb{P}^{\ell-t}_{z,y}(d\omega').$$

(A path chosen according to $\mathbb{P}^t_{x,z}$ is continued beyond time t with the constant value z; we thereby identify $\mathbb{P}^t_{x,z}$ with a law on $\mathcal{F}_{\ell-}$ that is concentrated on $\{\omega : \omega(s) = \omega(t-), t \le s < \ell\}$.) We write $\omega^* = (z, t, \omega, \omega')$ for the generic point of Ω^*, and $(\omega/t/\omega')$ for the path obtained by splicing ω and ω' together at time t. That is,

$$(\omega/t/\omega')(s) = \begin{cases} \omega(s), & \text{if } 0 \le s < t, \\ \omega'(s-t), & \text{if } s \ge t. \end{cases}$$

Notice that $\omega = (\omega/t/\theta_t\omega)$. Now define

$$Z(\omega^*) = z, \qquad T(\omega^*) = t,$$

$$Y_s(\omega^*) = X_s(\omega/t/\omega'), \quad 0 \le s < \ell,$$

$$A_s^*(\omega^*) = A_s(\omega/t/\omega'), \quad 0 \le s \le \ell.$$

Thus

$$\mathbb{P}^*(Z \in dz, T \in dt) = \rho(z,t)\nu(dz)\,dt,$$

and the conditional distribution of Y given $\{Z = z,\ T = t\}$ is the law $\mathbb{P}^t_{x,z} \circ \mathbb{P}^{\ell-t}_{z,y}$ on $(\Omega, \mathcal{F}_{\ell-})$ defined by

$$\begin{aligned} &\mathbb{P}^t_{x,z} \circ \mathbb{P}^{\ell-t}_{z,y}(H) \\ \text{(4.1)} \qquad &= \int_\Omega \int_\Omega \mathbb{P}^t_{x,z}(d\omega)\mathbb{P}^{\ell-t}_{z,y}(d\omega')H(\omega/t/\omega'). \end{aligned}$$

That is, under $\mathbb{P}^t_{x,z} \circ \mathbb{P}^{\ell-t}_{z,y}$ the path fragments $(Y_s)_{0\le s<t}$ and $(Y_{t+s})_{0\le s<\ell-t}$ are independent, with laws $\mathbb{P}^t_{x,z}$ and $\mathbb{P}^{\ell-t}_{z,y}$ respectively.

The following refinement of Proposition 3 is the key to our main result.

Lemma 2. *Let A be the CAF of X associated with ν as in section 3, and write*

$$\text{(4.2)} \qquad K = \int_0^\ell f(X_t, t)\, dA_t,$$

where f is a positive Borel function on $E\times]0,\ell[$. If $H \ge 0$ is $\mathcal{F}_{\ell-}$-measurable, then

(4.3)

$$\begin{aligned} &\mathbb{P}^\ell_{x,y}(HK) \\ &= \int_0^\ell dt \int_E \nu(dz) \left[\frac{p_t(x,z)p_{\ell-t}(z,y)}{p_\ell(x,y)}\right] f(z,t)\, \mathbb{P}^t_{x,z} \circ \mathbb{P}^{\ell-t}_{z,y}(H). \end{aligned}$$

Remark. Let ξ_ℓ be the random measure on $E\times]0,\ell[$ that is the image of dA_t (a measure on $]0,\ell[$) under the mapping $t \mapsto (X_t, t)$. the terminology of random measures, taken from Kallenberg [**K3**, Ch. 10] and reviewed in Section 5, the lemma states that under $\mathbb{P}^\ell_{x,y}$ the probabilities $\mathbb{P}^t_{x,z} \circ \mathbb{P}^{\ell-t}_{z,y}$ serve as the family of Palm distributions of ξ_ℓ. Just as this result follows from Proposition 3, it follows from Proposition 2 that for ξ the random measure on $E\times]0,\infty[$ that is the image of dA_t under the mapping $t \mapsto (X_t, t)$, under $\mathbb{P}_x$ the probabilities $\mathbb{P}^t_{x,z} \circ \mathbb{P}_z$ serve as the family of Palm distributions of ξ. In case A_t is the local time at a recurrent point in the state-space of X, the latter assertion is implicit in Kallenberg's work [**K1**].

Proof. Consider the process

$$J_t(z,\omega) = \int_\Omega \mathbb{P}^{\ell-t}_{z,y}(d\omega')H(\omega/t/\omega'), \quad 0 \le t < \ell.$$

It is easy to check that $((\omega,t),z) \mapsto J_t(z,\omega)$ is $\mathcal{P}_\ell \otimes \mathcal{E}$-measurable, where $\mathcal{P}_\ell$ is the predictable σ-algebra restricted to $\Omega \times [0,\ell[$ and $\mathcal{E}$ is the Borel σ-algebra on E. Moreover, $J_t(X_t(\omega),\omega)$ is a version of the $\mathbb{P}^\ell_{x,y}$-optional projection of the process $t \mapsto H$. (This is easy to see if H has the form $\Pi_k f_k(X_{t_k})$; an appeal to the monotone class theorem settles the matter.) Consequently,

$$\mathbb{P}^\ell_{x,y}(HK) = \mathbb{P}^\ell_{x,y}\left(\int_0^\ell Hf(X_t,t)\,dA_t\right)$$

$$= \mathbb{P}^{\ell}_{x,y}\left(\int_0^{\ell} J_t(X_t,\cdot) f(X_t,t)\, dA_t\right). \tag{4.4}$$

If we now apply Proposition 3 to the last term in (4.4), then (4.3) follows. □

We apply Lemma 2 with a specific choice of the function f in the definition (4.2) of K:

$$f(z,t) = \rho(z,t)\left[\frac{p_\ell(x,y)}{p_t(x,z)p_{\ell-t}(z,y)}\right]. \tag{4.5}$$

Proposition 4. *Let Z, T, and Y be defined on $(\Omega^*, \mathcal{F}^*, \mathbb{P}^*)$ as described above. If $H \geq 0$ is $\mathcal{F}_{\ell-}$-measurable, then*

$$\mathbb{P}^*(H(Y)) = \mathbb{P}^{\ell}_{x,y}(HK), \tag{4.6}$$

where $K = \int_0^{\ell} f(X_t,t)\, dA_t$ and f is given by (4.5). Moreover, the conditional distribution of (Z,T) given Y is determined by

$$\mathbb{P}^*(g(Z,T)\,|\,Y) = \frac{\int_0^{\ell} g(Y_t,t) f(Y_t,t)\, dA_t^*}{\int_0^{\ell} f(Y_t,t)\, dA_t^*}, \tag{4.7}$$

part of the assertion being that

$$\mathbb{P}^*\left(0 < \int_0^{\ell} f(Y_t,t)\, dA_t^* < \infty\right) = 1.$$

Proof. Let $H \geq 0$ be an $\mathcal{F}_{\ell-}$-measurable function on Ω, and let $g \geq 0$ be a Borel function on $E \times]0, \ell[$. Then by Lemma 2,

(4.8)
$$\begin{aligned}
&\mathbb{P}^*(H(Y)\, g(Z,T)) \\
&= \int_0^\ell dt \int_E \nu(dz) \rho(z,t)\, \mathbb{P}^t_{x,z} \circ \mathbb{P}^{\ell-t}_{z,y}(H)\, g(z,t) \\
&= \int_0^\ell dt \int_E \nu(dz) \\
&\qquad \left[\frac{p_t(x,z) p_{\ell-t}(z,y)}{p_\ell(x,y)} \right] f(z,t) \mathbb{P}^t_{x,z} \circ \mathbb{P}^{\ell-t}_{z,y}(H) g(z,t) \\
&= \mathbb{P}^\ell_{x,y}(H \tilde{K}),
\end{aligned}$$

where $\tilde{K} = \int_0^\ell g(X_t,t) f(X_t,t)\, dA_t$. The first assertion in Proposition 4 follows upon taking $g = 1$ in (4.8). Next, $A^*_s = A_s \circ Y$ is measurable over the $\mathbb{P}^*$-completion of $\sigma(Y)$, since A_s is measurable over the universal completion of $\mathcal{F}_s$. Let $K^* = \int_0^\ell f(Y_t,t)\, dA^*_t$, so that $K^* = K \circ Y$. By (4.8),

$$\mathbb{P}^*(K^* = 0) = \mathbb{P}^\ell_{x,y}(K; K = 0) = 0,$$

and

$$\mathbb{P}^*(K^* > \lambda) = \mathbb{P}^\ell_{x,y}(K; K > \lambda) \to 0 \text{ as } \lambda \to \infty,$$

since $\mathbb{P}^\ell_{x,y}(K) = \int_0^\ell dt \int_E \nu(dz)\, \rho(z,t) = 1$. Thus $\mathbb{P}^*(0 < K^* < \infty) = 1$. Now (4.7) follows easily from (4.8). □

Remark. It follows immediately from Proposition 4 that if $Q^* = (K^*)^{-1} \mathbb{P}^*$, then $Q^*(Y \in \cdot\,) = \mathbb{P}^\ell_{x,y}$, while $Q^*((Z,T) \in \cdot\,|Y) = \mathbb{P}^*((Z,T) \in \cdot\,|Y)$.

Example. Suppose that $x \in E$ is a regular point for X. That is, $\mathbb{P}_x(\inf\{t > 0 : X_t = x\} = 0) = 1$. Then by [**BG**, V,(3.13)] X admits local time at x. This is a CAF $L_t = L_t^x$ of X such that $L_t = \int_0^t 1_{\{x\}}(X_s)\, dL_s$ for all $t > 0$. The Revuz measure of L is proportional to the point mass at x, and we normalize L so that $\nu = \nu_L = \epsilon_x$. Thus, writing $p(s) = p_s(x,x)$,

$$\mathbb{P}_x(L_t) = \int_0^t p(s)\, ds, \qquad t > 0,$$

and

$$\mathbb{P}^\ell_{x,x}(L_t) = \int_0^t \left[\frac{p(s)p(\ell - s)}{p(\ell)}\right] ds, \qquad 0 < t \le \ell.$$

Let ρ be a probability density on $]0,\ell[$ and define $f(z,t) = f(t) = \rho(t)p(\ell)/[p(t)p(\ell - t)]$. Proposition 4 states that if we splice together two X-bridges (each beginning and ending at x) at an independent time T with law $\rho(t)\,dt$, then the law of the resulting process Y is absolutely continuous with respect to $\mathbb{P}^\ell_{x,x}$ with Radon-Nikodym derivative $\int_0^\ell f(t)\, dL_t$. In particular, the choice $\rho(t) = [p(t)p(\ell - t)]/[p(\ell)c(\ell)]$, where $c(\ell) = \mathbb{P}^\ell_{x,x}(L_\ell)$, leads to $f \equiv 1/c(\ell)$, so the density factor is simply $L_\ell/c(\ell)$. In case X is one-dimensional Brownian motion, these are the results presented in the introduction around formulae (1.0) and (1.1).

There is a multivariate version of the result of this section, whose formulation and proof we leave to the reader. The idea is to replace (Z,T) by $(Z_1,T_1),\ldots,(Z_n,T_n)$, where $T_1 < T_2 < \cdots < T_n$. By making a suitable choice of the joint density of $(Z_1,T_1),\ldots,(Z_n,T_n)$ one can obtain a spliced process whose density relative to the X-bridge is a multiple of $(A_\ell)^n/n!$.

Consider for example the setup in the introduction, with B a Brownian bridge on $[0,1]$, U uniformly distributed over $[0,1]$, independent of B, and let $]G,D[$ be the excursion interval straddling U. Using the tripartite decomposition of B at times G and D into bridge/excursion/bridge, and the fact that given G and D, U is uniform on $]G,D[$, we can use U to perform the (inverse) Vervaat shuffle on the excursion fragment [**V,Bi,BP**], thereby obtaining three adjacent bridges. The joint law of G and D is then exactly what is required to make the transformed process have unconditional law with density a constant times $(L_1)^2$ relative to the standard bridge. Compare with [**PPY**, Corollary 3.15]: a process whose law has density a constant times $(L_1)^n$ relative to that of the standard bridge B is obtained from B by length biased sampling without replacement of n excursion intervals, closing up the gaps, and standardizing.

5. A General Probabilistic Interpretation of Palm Measures

The passage from Lemma 2 to Proposition 4 illustrates a general calculation involving random measures and their Palm measures which we have found useful in other contexts (see [**PPY**, Lemma 4.1] [**PY2**, Lemma 2.2]), and which it seems worth recording in a general setting. The general result of this calculation appears in Proposition 5 below.

Let $(\Omega, \mathcal{F}, Q)$ be a probability space, $(S, \mathcal{S})$ a measurable space, and ξ a random measure on S based on $(\Omega, \mathcal{F}, Q)$. That is, $\xi(w, A)$ is a $\mathcal{F}$-measurable function of $w \in \Omega$ ($A \in \mathcal{B}(S)$ fixed) and a measure on $(S, \mathcal{B}(S))$ as a function of A ($w \in \Omega$ fixed). The intensity measure of ξ is

$$(5.1), \qquad \lambda(A) = \int \xi(w, A)\, Q(dw), \qquad A \in \mathcal{B}(S),$$

which we assume to be σ-finite. Let us assume that the bimeasure

$$Q(\xi(A); B), \qquad A \in \mathcal{B}(S),\ B \in \mathcal{F},$$

admits a disintegration

$$(5.2) \qquad Q(\xi(A); B) = \int_A \lambda(dx)\, Q^x(B),$$

where $(x, B) \mapsto Q^x(B)$ is a Markov kernel from $(S, \mathcal{B}(S))$ to $(\Omega, \mathcal{F})$. (For a disintegration as in (5.2) to exist, it is sufficient that Ω be a Polish space and that $\mathcal{F}$ be the corresponding Borel σ-algebra.)

The collection $\{Q^x : x \in S\}$ is the family of *Palm distributions* associated with the random measure ξ. When ξ is the random counting measure corresponding to a point process with no multiple points, Q^x may be interpreted intuitively as Q conditioned on ξ putting a point at x. See for example [**K2**]. In case ξ is a more general random measure, in particular if ξ is diffuse, the intuitive meaning of the Palm measures is less clear. Still, the following proposition offers a general probabilistic interpretation of Palm measures in terms of expanding the probability space to allow that, given ξ, a point $X \in S$ is picked at random with some density relative to ξ.

Proposition 5. *Suppose we have defined on a probability space* $(\Omega^*, \mathcal{F}^*, P^*)$ *an* S*-valued random variable* X *and an* Ω*-valued random variable* W. *If*

$$P^*(X \in dx) = f(x)\lambda(dx) \tag{5.3}$$

for some probability density f *on* S *relative to the intensity measure* λ, *and*

$$P^*(W \in dw | X = x) = g(w|x)Q^x(dw) \tag{5.4}$$

for some jointly measurable $g(w|x)$, *then*

$$P^*(W \in dw) = h(w)Q(dw), \tag{5.5}$$

$$(5.6) \qquad P^*(X \in dx | W = w) = j(x|w)\xi(w, dx),$$

where

$$(5.7) \qquad h(w) = \int_S f(x)g(w|x)\xi(w, dx),$$

$$(5.8) \qquad j(x|w) = f(x)g(w|x)/h(w).$$

Conversely, if the joint law of W and X is such that (5.5) and (5.6) hold for some probability density h relative to Q, and some jointly measurable $j(x|w)$, then (5.3) and (5.4) hold with

$$(5.9) \qquad f(x) = \int_\Omega h(w)j(x|w)Q^x(dw),$$

$$(5.10) \qquad g(w|x) = h(w)j(x|w)/f(x).$$

The proof just uses the definition of the Palm distributions: $Q(dw)\xi(w, dx) = \lambda(dx)Q^x(dw)$, and a standard Bayes type calculation, as spelled out for instance in [**L**, Lemma 2.1].

In conclusion, we reflect that in recent years arguments involving random measures and their associated Palm measures have played an increasingly important role in the theory of Markov processes. See for instance [**AM, Fi, G**]. Just as bridge laws can be interpreted as Palm distributions, so can excursion laws. See for instance [**Pi**], where a generalization of Bismut's [**B**] decomposition of the Brownian excursion is explained in terms of Palm distributions.

References

[AP] D. Aldous and J. Pitman: Brownian bridge asymptotics for random mappings. Dept. Stat., U.C. Berkeley, Tech. Report No. 369, 1992.

[AM] B.W. Atkinson and J.B. Mitro: Applications of Revuz and Palm type measures for additive functionals in weak duality. *Seminar on Stochastic Processes 1982*, Birkhäuser, Boston, (1983) 23–50.

[A] J. Azéma: Théorie générale des processus et retournement du temps. *Ann. Scient. de l'École Norm. Sup.*, 4e série, **6** (1973) 459–519.

[AJ] J. Azéma and T. Jeulin: Précisions sur la mesure de Föllmer. *Ann. Inst. Henri Poincaré*, **12** (1976) 257–283.

[BP] J. Bertoin and J. Pitman: Path transformations connecting Brownian bridge, excursion and meander. To appear in *Bull. Soc. Math. France.*

[Bi] P. Biane: Relations entre pont et excursion du mouvement brownien réel. *Ann. Inst. Henri Poincaré*, **22** (1986) 1-7.

[B] J.M. Bismut: Last exit decomposition and regularity at the boundary of transition probabilities. *Zeitschrift für Wahrscheinlichkeitstheorie verw. Gebiete.* **69** (1985) 65–98.

[BG] R. M. Blumenthal and R. K. Getoor: *Markov Processes and Potential Theory.* Academic Press, New York, 1968.

[D] E. B. Dynkin: Minimal excessive measures and functions. *Trans. Amer. Math. Soc.* **258** (1980) 217–244.

[EL] M. Emery and R. Léandre: Sur une formule de Bismut. Sém. Prob. XXIV, Lecture Notes in Math. **1426**, Springer, 1990, pp. 448–452.

[Fi] P.J. Fitzsimmons: Homogeneous random measures and a weak order for the excessive measures of a Markov process. *Trans. Amer. Math. Soc.* **303** (1987) 421–478.

[F] H. Föllmer: The exit measure of a supermartingale. *Zeitschrift für Wahrscheinlichkeitstheorie verw. Gebiete.* **21** (1972) 154–166.

[GS1] R. K. Getoor and M. J. Sharpe: Some random time dilations of a Markov process *Math. Z.* **167** (1979) 187–199.

[GS2] R. K. Getoor and M. J. Sharpe: The Markov property at cooptional times. *Zeitschrift für Warscheinlichkeitstheorie verw. Gebiete.* **48** (1979) 201–211.

[GS3] R. K. Getoor and M. J. Sharpe: Markov properties of a Markov process. *Zeitschrift für Warscheinlichkeitstheorie verw. Gebiete.* **55** (1981) 313–330.

[GS4] R. K. Getoor and M. J. Sharpe: Excursions of dual processes. *Adv. Math.* **45** (1982) 259–309.

[G] R. K. Getoor: *Excessive measures.* Birkhäuser, Boston, 1990.

[H] P. Hsu: Brownian bridges on Riemannian manifolds. *Probability Theory and Related Fields.* **84** (1990) 103–118.

[IM] K. Itô and H. P. McKean: *Diffusion processes and their sample paths.* Springer, Berlin, 1965.

[JY] T. Jeulin and M. Yor. Sur les distributions de certaines fonctionelles du mouvement brownien. Sém. Prob. XV, Lecture Notes in Math. **850**, Springer, 1981, pp. 210-226.

[K1] O. Kallenberg: Splitting at backward times in regenerative sets. *Ann. Probab.* **9** (1981) 781–799.

[K2] O. Kallenberg: An informal guide to the theory of conditioning in point processes. *Int. Statist. Review* **52** (1984), 151–164.

[K3] O. Kallenberg: *Random Measures*, 4th ed., Akademie-Verlag and Academic Press, Berlin and London 1986. (1st edition 1975).

[L] G. Last: Some remarks on conditional distributions for point processes. *Stoch. Proc. Appl.*. **34** (1990) 121–135.

[M1] P.-A. Meyer: Processus de Markov: La frontière de Martin. Lecture Notes in Math. **77**, Springer, 1970.

[M2] P.-A. Meyer: Note sur l'interprétation des mesures d'équilibre. Sém. Prob. VII, Lecture Notes in Math. **321**, Springer, 1973, pp. 210–217.

[Pa] K.R. Parthasarathy: *Probability Measures on Metric Spaces.* Academic Press, New York-London, 1967.

[PPY] Perman, M., Pitman, J., and Yor, M. Size-biased sampling of Poisson processes and excursions. *Probability and Related Fields.* **92** (1992) 21-39.

[Pi] J. Pitman: Stationary excursions. Sém. Prob. XXI, Lecture Notes in Math. **1247**, Springer, 1987, pp. 289–302.

[PY1] J. Pitman and M. Yor: A decomposition of Bessel bridges. *Zeitschrift für Wahrscheinlichkeitstheorie verw. Gebiete.* **59** (1982) 425–458.

[PY2] J. Pitman and M. Yor: Arcsine laws and interval partitions derived from a stable subordinator. *Proc. London Math. Soc. (3).* **65** (1992) 326–356.

[R] D. Revuz: Mesures associées aux fonctionnelles additives de Markov, I. *Trans. Amer. Math. Soc.* **148** (1970) 501–531.

[RY] D. Revuz and M. Yor: *Continuous martingales and Brownian motion.* Springer, Berlin-Heidelberg, 1991.

[Sa] P. Salminen: Brownian excursions revisited. *Seminar on Stochastic Processes 1983*, Birkhäuser, Boston, (1984) 161–187.

[SW] G. R. Shorack and J. A. Wellner: *Empirical Processes with Applications to Statistics.* Wiley, 1986.

[V] W. Vervaat: A relation between Brownian bridge and Brownian excursion. *Ann. Probab.* **7** (1979) 143–149.

[Wa] J. B. Walsh: The perfection of multiplicative functionals. Sém. Prob. VI, Lecture Notes in Math. **258**, Springer, 1972, 233–242.

[Wi] R. Wittmann: Natural densities of Markov transition probabilities. *Prob. Theor. Rel. Fields.* **73** (1986) 1–10.

[Y] J.-A. Yan: A formula for densities of transition functions. Sém. Prob. XXII, Lecture Notes in Math. **1321**, Springer, 1988, pp. 92–100.

[Yo] M. Yor: A propos de l'inverse du mouvement brownien dans $\mathbb{R}^n$ $(n \geq 3)$. *Ann. Inst. Henri Poincaré*, **21** (1984) 27-38.

Pat FITZSIMMONS
Department of Mathematics
University of California, San Diego
9500 Gilman Drive
La Jolla, California 92093–0112

Jim PITMAN
Department of Statistics
University of California
Berkeley, California 94720

Marc YOR
Laboratoire de Probabilités
Université Pierre et Marie Curie
4, Place Jussieu–Tour 56
75252 Paris Cedex 05
France

The Martingale Problem for a Differential Operator with Piecewise Continuous Coefficients

Ping Gao

0. Introduction

Let L be a linear second order elliptic differential operator defined by

$$Lu = \sum_{i,j=1}^{n} a_{i,j}(x)\frac{\partial^2 u}{\partial x_i \partial x_j} + \sum_{j=1}^{n} b_j(x)\frac{\partial u}{\partial x_j} \tag{1}$$

with $b = (b_j)$, $a = (a_{ij})$ bounded and measurable, and a symmetric. Suppose a is uniformly positive definite, i.e. there exist positive numbers μ and ν such that for all $y = (y_1, \ldots, y_n)$, and every $x \in \mathbb{R}^n$

$$\mu \sum_{i=1}^{n} y_i^2 \le \sum_{i,j=1}^{n} a_{ij}(x) y_i y_j \le \nu \sum_{i=1}^{n} y_i^2 . \tag{2}$$

We shall sometimes denote L by $L_{(a,b)}$ to indicate the relation between L and a, b. Let $W(\mathbb{R}^n) = C([0,\infty), \mathbb{R}^n)$, and $\mathcal{B}(W(\mathbb{R}^n))$, $\mathcal{B}_t(W(R^n))$ be the σ-field generated by $\{\omega(t) : \omega(\cdot) \in W(\mathbb{R}^n), 0 \le t < \infty\}$ and $\{\omega(s) : \omega(\cdot) \in W(\mathbb{R}^n), 0 \le s < t\}$ respectively.

Given $x \in \mathbb{R}^n$, a solution to the martingale problem for (a, b) starting from x is a probability measure P_x on $(W(\mathbb{R}^n), \mathcal{B}(W(\mathbb{R}^n))$ satisfying

(i) $P_x\{\omega : \omega(0) = x\} = 1$;

(ii) $f(\omega(t)) - f(\omega(0)) - \int_0^t (Lf)(\omega(s))ds$

is a $(P_x, \mathcal{B}_t(W(\mathbb{R}^n)))$-martingale for every $f \in C_k^2(\mathbb{R}^n)$.

If for each $x \in \mathbb{R}^n$, there is exactly one such solution, then the martingale problem for (a,b) is well posed (see [10]).If the family of probabilities $\{P_x, x \in \mathbb{R}^n\}$ is a strong Markovian system, then $\{P_x, x \in \mathbb{R}^n\}$ is called an $L_{(a,b)}$-diffusion, or a diffusion generated by $L_{(a,b)}$ (see [5]). If the martingale problem is well posed, then the family $\{P_x, x \in \mathbb{R}^n\}$ is a strong Markovian system (see [5]). On the other hand, if a, b are bounded and measurable with a being uniformly positive definite, then an $L_{(a,b)}$- diffusion always exists (see [10],exercise 12.4.3).When $n = 1, 2$, such an $L_{(a,b)}$-diffusion is unique (see [10], exercise 7.3.3 & 7.3.4). When $n \geq 3$, if $a = (a_{ij})$ is continuous, then uniqueness holds([5],[10]).If a is discontinuous, the question of uniqueness remains open (see [3],[6]).

In this paper, we show that uniqueness holds (i.e. the martingale problem for (a,b) is well posed) if $a = (a_{ij})$ is piecewise continuous on each of the two half spaces. i.e. if

$$a_{ij}(x) = \begin{cases} A_{ij}^{+}(x) & \text{if } x \in \mathbb{R}_{+}^{n} = \{x : x_1 > 0\} \\ A_{ij}^{-}(x) & \text{if } x \in \mathbb{R}_{-}^{n} = \{x : x_1 < 0\} \end{cases} \tag{3}$$

where $A_{ij}^{+}(x)$, $A_{ij}^{-}(x)$ are continuous on $\overline{\mathbb{R}_{+}^{n}} = \{x : x_1 \geq 0\}$ and $\overline{\mathbb{R}_{-}^{n}} = \{x : x_1 \leq 0\}$, respectively.

After we obtained this result, we learned of a preprint of P.L. Lions about 12 years old in which a similar result is proved; it does not seem to have ever been published. Our proof uses a result of A. Lorenzi [9] and the technique of Stroock and Varadhan, and should be of independent interest.

1. The Uniqueness of the Solution to the Martingale Problem

Let L be as in (1), where $b_j \in L^\infty(\mathbb{R}^n)$, $a_{ij}(x) = a_{ji}(x)$, and also a satisfies (2),(3).

Theorem 1.1 *Let $a = (a_{ij})$, $b = (b_j)$ be as described above. Then the martingale problem for (a,b) is well posed.*

Proof : (i) As we have mentioned, for each $x \in \mathbb{R}^n$, the martingale problem for $L_{(a,b)}$ starting at x has at least one solution P_x . We will show uniqueness, i.e. we show that if P_x, and P_x' are two solutions, then $P_x = P_x'$ on $W(\mathbb{R}^n)$.

Let $\Omega = W(\mathbb{R}^n)$. For each $\omega \in \Omega$, define $X_t(\omega) = X(t,\omega) = \omega(t) = X(t)$. Since the n- dimensional Lebesgue measure of $\{x : x_1 = 0\}$ is zero, and $a = (a_{ij})$ is uniformly positive definite, the process $X(t)$ spends zero time at $\{x : x_1 = 0\}$ under each P_x, or P_x' for every $x \in \mathbb{R}^n$(see [7]). Therefore the value of $(a_{ij}(x))$ at $\{x : x_1 = 0\}$ does not affect P_x, or P_x', $(\forall x \in \mathbb{R}^n)$. For convenience, we assume $a_{ij}(x) = A_{ij}^{+}(x)$, for $x \in \overline{\mathbb{R}_{+}^{n}} = \{x :$

$x_1 \geq 0\}$.Also, without loss of generality, we assume $b_j(x) \equiv 0$, $j = 1, \ldots, n$, since the general case can be reduced to this case by the transformation of drift(see [5], also see [10], theorem 6.4.3). Let E_x, E'_x denote the expectation with respect to P_x, P'_x respectively.

Fix $\lambda > 0$, for each $f \in C_b^2(\mathbb{R}^n)$, we define

$$V_\lambda f(x) = E_x \int_0^\infty e^{-\lambda t} f(X(t,\omega))dt.$$

We define $V'_\lambda f(x)$ similarly.

In order to show $P_x = P'_x$, we only need to show $V_\lambda f(x) = V'_\lambda f(x)$ for every $f \in C_k^2(\mathbb{R}^n)$ and every $x \in \mathbb{R}^n$(see [5], p 192).

If $A_{ij}^+(x) \equiv \tilde{A}_{ij}^+ =$ constant, $A_{ij}^-(x) \equiv \tilde{A}_{ij}^- =$ constant, we denote the corresponding a as $\tilde{a} = (\tilde{a}_{ij})$, i.e.

$$\tilde{a}_{ij}(x) = \begin{cases} \tilde{A}_{ij}^+ = \text{constant} & x_1 \geq 0 \\ \tilde{A}_{ij}^- = \text{constant} & x_1 < 0. \end{cases}$$

In this case , we say $\tilde{a}$ is piecewise constant, and denote $L_{(\tilde{a},0)}$ as $\tilde{L}$, and $L_{(\tilde{a},0)} - \lambda$ as $\tilde{L}_\lambda$. We also define $\tilde{P}_x$, $\tilde{E}_x$, $\tilde{V}_\lambda$ accordingly.

(ii) Let's first look at the properties of the operator $\tilde{V}_\lambda$.

If $u \in C_b^2(\mathbb{R}^n)$, then by Itô's formula,

$$\tilde{E}_x u(X(t)) = u(x) + \int_0^t \tilde{E}_x \tilde{L} u(X(s))ds.$$

For fixed $\lambda > 0$, apply Fubini's theorem:

$$\tilde{E}_x \int_0^\infty e^{-\lambda t} \lambda u(X(t))dt = u(x) + \tilde{E}_x \int_0^\infty e^{-\lambda t} \tilde{L} u(X(t))dt.$$

or,

$$\tilde{V}_\lambda(\tilde{L}_\lambda u)(x) = -u(x). \tag{4}$$

A result of Lorenzi ([9]) states the following: with $\tilde{L}_\lambda$ as above,

(1.1) for each $1 < p < \infty$, there exists a unique operator $\tilde{T}_\lambda : L^p(\mathbb{R}^n) \to W^{2,p}(\mathbb{R}^n)$, such that for each $h \in L^p(\mathbb{R}^n)$,

$$||\tilde{T}_\lambda h||_{2,p,\mathbb{R}^n} \leq C_{(p,\lambda,\mu,\nu)} ||h||_{p,\mathbb{R}^n}, \tag{5}$$

$$\tilde{L}_\lambda(\tilde{T}_\lambda h) = h, \tag{6}$$

where $C = C_{(p,\lambda,\mu,\nu)}$, which only depends on p, λ, μ, and ν. And for $u \in W^{2,p}(\mathbb{R}^n)$,

$$\tilde{T}_\lambda(\tilde{L}_\lambda u) = u.$$

By (4), notice for $u \in C_k^\infty(\mathbb{R}^n) \subset C_b^2(\mathbb{R}^n) \subset W^{2,p}(\mathbb{R}^n)$,(where $C_k^\infty(\mathbb{R}^n) = \{h : h \in C^\infty(\mathbb{R}^n), h \text{ has compact support}\}$), we have

$$\tilde{V}_\lambda = -\tilde{T}_\lambda \qquad \text{on} \qquad \tilde{L}_\lambda(C_k^\infty(\mathbb{R}^n))$$

and

$$||\tilde{V}_\lambda(g)||_{2,p,\mathbb{R}^n} \leq C||g||_{p,\mathbb{R}^n} \quad \text{for} \quad g \in \tilde{L}_\lambda(C_k^\infty(\mathbb{R}^n)).$$

Now $C_k^\infty(\mathbb{R}^n)$ is dense in $W^{2,p}(\mathbb{R}^n)$ (see [1], Corollary 3.19). For each $u \in W^{2,p}(\mathbb{R}^n)$, we choose $u_m \in C_k^\infty(\mathbb{R}^n)$, so that $||u_m - u||_{2,p,\mathbb{R}^n} \to 0, (m \to \infty)$, and then

$$\tilde{V}_\lambda(\tilde{L}_\lambda u_m)(x) = -u_m(x).$$

Since $W^{2,p}(\mathbb{R}^n) \subset C^1(\mathbb{R}^n)$,(see [4] or [1]), for every $x \in \mathbb{R}^n$, $u_m(x) \to u(x)$, $(m \to \infty)$. Hence

$$\tilde{V}_\lambda(\tilde{L}_\lambda u_m)(x) \to -u(x), \quad \forall x \in \mathbb{R}^n.$$

By [8] Theorem 2, if $p \geq n+1$, then there exists a constant N, such that

$$|\tilde{V}_\lambda(g)(x)| \leq N||g||_{p,\mathbb{R}^n}, \qquad \forall g \in L^\infty(\mathbb{R}^n).$$

Therefore

$$\begin{aligned} |\tilde{V}_\lambda(\tilde{L}_\lambda u - \tilde{L}_\lambda u_m)(x)| &\leq N||\tilde{L}_\lambda u - \tilde{L}_\lambda u_m||_{p,\mathbb{R}^n} \\ &\leq N'||u - u_m||_{2,p,\mathbb{R}^n} \to 0 \quad (m \to \infty). \end{aligned}$$

Hence

$$\tilde{V}_\lambda(\tilde{L}_\lambda u)(x) = -u(x).$$

Thus we have proved , for each $u \in W^{2,p}(\mathbb{R}^n)$,

$$\tilde{V}_\lambda(\tilde{L}_\lambda u)(x) = -u(x). \tag{7}$$

Since $\tilde{L}_\lambda : W^{2,p}(\mathbb{R}^n) \longrightarrow L^p(\mathbb{R}^n)$ is onto([9], see (1.1) above), we have, for $f \in L^p(\mathbb{R}^n)$, $p \geq n+1$,

$$\tilde{V}_\lambda(f) = -\tilde{T}_\lambda f \tag{8}$$

and

$$||\tilde{V}_\lambda(f)||_{2,p,\mathbb{R}^n} \leq C||f||_{p,\mathbb{R}^n} \tag{9}$$

where C depends only on λ, p, μ, ν. In particular, if $f \in C_k(\mathbb{R}^n)$, and $p \geq n+1$, then $f \in L^p(\mathbb{R}^n)$, and $\tilde{V}_\lambda(f) \in W^{2,p}(\mathbb{R}^n)$,

$$||\tilde{V}_\lambda(f)||_{2,p,\mathbb{R}^n} \leq C||f||_{p,\mathbb{R}^n}, \tag{10}$$

$$\forall i,j \leq n, \qquad ||\frac{\partial^2 \tilde{V}_\lambda f}{\partial x_i \partial x_j}||_{p,\mathbb{R}^n} = ||D_{ij}(\tilde{V}_\lambda f)||_{p,\mathbb{R}^n} \leq C||f||_{p,\mathbb{R}^n} \tag{11}$$

(iii) Next let $\tilde{a}$ be fixed, and let $0 < \varepsilon_0 < (n^2 C_{(p,\lambda,\mu,\nu)})^{-1}$. We will show if $\hat{a} = \{\hat{a}_{ij}(x)\}$ is a matrix satisfying the conditions of the theorem, and

$$\sup_{x \in \mathbb{R}^n} |\hat{a}_{ij}(x) - \tilde{a}_{ij}| \leq \varepsilon_0,$$

then the solution to the martingale problem for $\hat{L} \stackrel{\text{def}}{=} L_{(\hat{a},0)}$ starting at x is unique for every $x \in \mathbb{R}^n$.

Let $\hat{P}_x$, $\hat{P}'_x$ be two such solutions. We need to show $\hat{V}_\lambda f(x) = \hat{V}_\lambda f(x)$ for each $f \in C_k^2(\mathbb{R}^n) \subset C_b^2(\mathbb{R}^n)$. As in (4), we apply [8] Theorem 2 to $\hat{V}_\lambda$, $\hat{V}'_\lambda$, and we have, for $p \geq n+1$,

$$\sup_{||f||_{p,\mathbb{R}^n} \leq 1, x \in \mathbb{R}^n} |(\hat{V}_\lambda - \hat{V}'_\lambda) f(x)| \leq 2N \quad < \infty. \tag{12}$$

By (4) , as in the proof of (7), for $f \in W^{2,p}(\mathbb{R}^n)$, we have

$$\hat{V}_\lambda(\lambda f - \tilde{L} f)(x) = f(x) + \hat{V}_\lambda(\hat{L} f - \tilde{L} f)(x). \tag{13}$$

Let $h \in C_k^2(\mathbb{R}^n)$. Then by (10) and (8),

$$f \stackrel{\text{def}}{=} \tilde{V}_\lambda h \in W^{2,p}(\mathbb{R}^n), \qquad \tilde{L}_\lambda f = -h.$$

Let

$$c_{ij}(x) = \hat{a}_{ij}(x) - \tilde{a}_{ij} .$$

Then (13) becomes

$$\hat{V}_\lambda(h)(x) = \tilde{V}_\lambda h(x) + \hat{V}_\lambda(K_\lambda h)(x),$$

where

$$K_\lambda h(y) = \sum_{i,j=1}^{n} c_{ij}(y) D_{ij} \tilde{V}_\lambda h(y).$$

We have a similar result for $\hat{P}'_x$. Hence

$$(\hat{V}_\lambda - \hat{V}'_\lambda) h(x) = (\hat{V}_\lambda - \hat{V}'_\lambda)(K_\lambda h)(x).$$

Since $|c_{ij}(\cdot)| \leq \varepsilon_0$,

$$\begin{aligned} \sup_{\|h\|_{p,\mathbb{R}^n}\leq 1} |(\hat{V}_\lambda - \hat{V}'_\lambda)h(x)| &= \sup_{\|h\|_{p,\mathbb{R}^n}\leq 1} |(\hat{V}_\lambda - \hat{V}'_\lambda)(K_\lambda h)(x)| \\ &\leq \sup_{\|f\|_{p,\mathbb{R}^n}\leq 1} |(\hat{V}_\lambda - \hat{V}'_\lambda)f(x)| \sup_{\|h\|_{p,\mathbb{R}^n}\leq 1} \|K_\lambda h\|_{p,\mathbb{R}^n} \\ &\leq \sup_{\|f\|_{p,\mathbb{R}^n}\leq 1} |(\hat{V}_\lambda - \hat{V}'_\lambda)f(x)| n^2 \varepsilon_0 C_{(p,\lambda,\mu,\nu)}. \end{aligned}$$

Using (12),

$$(1 - n^2\varepsilon_0 C) \sup_{\|h\|_{p,\mathbb{R}^n}\leq 1} |(\hat{V}_\lambda - \hat{V}'_\lambda)h(x)| \leq 0.$$

Since $1 - n^2\varepsilon_0 C_{(p,\lambda,\mu,\nu)} > 0$, we must have

$$\sup_{\|h\|_{p,\mathbb{R}^n}\leq 1} |(\hat{V}_\lambda - \hat{V}'_\lambda)h(x)| = 0.$$

Therefore $\hat{V}_\lambda h(x) = \hat{V}'_\lambda h(x)$, and we conclude that $\hat{P}_x = \hat{P}'_x$ on Ω.

(iv) Let $a = (a_{ij}(x))$ satisfy the conditions of the theorem. We can show that for each $\overline{x} \in \mathbb{R}^n$, there exist a piecewise constant matrix $\tilde{a}$, an open set $\mathcal{O}_{\overline{x}} \ni \overline{x}$, and a piecewise constant matrix $\hat{a}$ such that $\hat{a} = a$ on $\mathcal{O}_{\overline{x}}$, and $\hat{a}$ satisfies the conditions of (iii), so the martingale problem for $\hat{a}$ is well posed. Then by applying [10] Theorem 6.6.1, we conclude that the martingale problem for a is well posed.

Acknowledgement : This is a part of the author's Ph.D dissertation. I would like to sincerely thank my academic adviser, Professor R. F. Bass , for his guidance and help.

References

[1] **R.A. Adams**: *Sobolev Spaces.*New York: Academic Press,1975.

[2] **R.F. Bass**: *The Dirichlet Problem For Radially Homogeneous Elliptic Operators.* Trans. of the American Math. Society. Vol.320. no.2. Aug.1990. p. 593-614.

[3] **R.F. Bass and E. Pardoux**: *Uniqueness for Diffusions with Piecewise Constant Coefficients.* Probability Theory and Related Fields. 76. (1987),557-572.

[4] **D. Gilbarg, N.S.Trudinger**: *Elliptic Partial Differential Equations of Second Order.* 2nd.ed. Springer. New York 1983.

[5] **N. Ikeda, S. Watanabe**: *Stochastic Differential Equations and Diffusion Processes,* Tokyo: North Holland/ Kodansha, 1981.

[6] **N.V. Krylov**: *On Itô's Stochastic Integral Equations,* Theory Prob. Applications, 14 (1969) 330-336.

[7] **N.V. Krylov**: *An Inequality in the Theory of Stochastic Integrals.* Theor. Probab. Appl. 16, (1971) 438-448.

[8] **N.V. Krylov**: *Some Estimates of the Probability Density of a Stochastic Integral.* Math. USSR.Izvestija, Vol.8 (1974). No.1

[9] **A. Lorenzi**: *On Elliptic Equations with Piecewise Constant Coefficients II*. Ann. Scuola, Norm. Sup. Pisa (3) 26 (1972) 839-870.

[10] **D.W. Stroock,S.R.S. Varadhan**: *Multidimensional Diffusion Processes.* Springer, New York. 1979.

GENERALIZED SCHRÖDINGER SEMIGROUPS

J. GLOVER*, M. RAO AND R. SONG

0. INTRODUCTION

Let $X = (X_t)$ be a standard Brownian motion on R^d and let μ be a Radon measure in the Kato class K_d. (See (1.1) below for the definition of the Kato class.) If $A = (A_t)_{t>0}$ is the continuous additive functional of X associated with μ, then the semigroup $(S_t)_{t>0}$ defined below

$$S_t f(x) = E^x \left\{ e^{A_t} f(X_t) \right\}, \qquad x \in R^d$$

is called a Schrödinger semigroup and it is of great importance in physics, and thus it has been studied by many authors, see [3], [13] and the references therein. Among other things, it is proved that the Schrödinger semigroup $(S_t)_{t>0}$ admits a jointly continuous symmetric integral kernel $q(t, x, y)$ with the property that there exist constants C and β such that for any $t > 0$ and any $x, y \in R^d$,

(0.1)

$$C^{-1} e^{-\beta t} t^{-\frac{d}{2}} \exp\left(\frac{-3|x-y|^2}{4t}\right) \le q(t,x,y) \le C e^{\beta t} t^{-\frac{d}{2}} \exp\left(\frac{-|x-y|^2}{4t}\right).$$

The assumption that μ is in the Kato class implies that the additive functional (A_t) with Revuz measure μ is of bounded variation, which is crucial in all the arguments in the references mentioned above. However, there are a lot of naturally occuring additive functionals which are not of bounded variation. How can these be investigated, and can we define a semigroup similar to $(S_t)_{t>0}$? If such a semigroup can be defined, does this new semigroup also have a jointly continuous symmetric integral kernel $q(t, x, y)$ satisfying (0.1)?

* Research of the first author supported in part by NSA Grant MDA-92-H-30324

The study of the questions of above was initiated in [10] and a class of continuous additive functionals not necessarily of bounded variation was introduced and studied there. A bounded continuous function $u \in H^1(R^d)$ is said to be admissible if $|\nabla u|^2$ is in the Kato class. For an admissible function u,

$$A_t^{[u]} = u(X_t) - u(X_0) - \int_0^t \nabla u(X_s) \cdot dX_s$$

is a continuous additive functional of zero energy with respect to X, and we call $A_t^{[u]}$ the continuous additive functional generated by u. In general, $A_t^{[u]}$ is not of bounded variation. It was proved in [10] that if $\mu \in K_d$ is of compact support and if (A_t) is the continuous additive functional associated with μ, then there exists an admissible u such that (A_t) is generated by u. Thus the class of additive functionals generated by admissible functions partially generalizes the class of additive functionals associated with measures in the Kato class. A continuous additive functional (A_t) is said to be special if it can be decomposed as the sum of two additive functionals $(\tilde{A}_t)$ and $(\overline{A}_t)$, with $(\tilde{A}_t)$ being generated by an admissible function and $(\overline{A}_t)$ being associated with a measure in the Kato class. Generally the decompsition above is not unique. For a special additive fuctional (A_t) with $A_t = A_t^{[u]} + \overline{A}_t$, we can define

$$T_t f(x) = E^x \left\{ e^{A_t} f(X_t) \right\}, \qquad x \in R^d.$$

Since the class of special additive functionals generalizes the class of additive functionals associated with measures in the Kato class, it is reasonable to call $(T_t)_{t>0}$ a generalized Schrödinger semigroup.

The purpose of this paper to prove that the generalized Schrödinger semigroup $(T_t)_{t>0}$ also admits a jointly continuous symmetric integral kernel $q(t,x,y)$ satisfying (0.1). (See Theorem (2.8) and Theorem (2.9) below.) We are also going to show that if D is a connected open subset of R^d and τ_D is the first exit time of X_t from D, then the generalized Schrödinger semigroup $(T_t^D)_{t>0}$ on D defined below

$$T_t^D f(x) = E^x \left\{ e^{A_t} f(X_t); t < \tau_D \right\}, \qquad x \in D,$$

admits a strictly positive, jointly continuous symmetric integral kernel $q^D(t,x,y)$ satisfying the right half of (0.1). (See Theorem (2.11) below.) Once Theorem (2.11) is proven, we can follow a routine argument which has been used in [3], [11] and [12] to give a new proof of the gauge theorem in [10]. This will be done in Section 3. Compared with the proof given in [10], the new proof heavily depends on the symmetry of X_t. But the new proof has an advantage: it can be easily extended to prove a generalization of the gauge theorem for the mixed boundary value problem in [12] once we can prove that the corresponding generalized Schrödinger semigroup admits a nice integral kernel.

Having seen the results of this paper, one would naturally ask the following question: What is the infinitesimal generator of S_t? or equivalently, what is quadratic form corresponding to S_t? This question is far from easy and we will deal with it in a subsequent paper.

The strategy of this paper is to construct a sequence of infinitely differentiable functions u_n to approximate the admissible function u. Since the u_n's we are going to construct are infinitely differentiable functions with compact support, we know that

$$A_t^{[u_n]} = \frac{1}{2}\int_0^t \Delta u_n(X_s)ds,$$

therefore for each $n \geq 1$, the semigroup $(T_t^{(n)})_{t>0}$ defined below

$$T_t^{(n)} f(x) = E^x\{\exp(A_t^{[u_n]} + \overline{A}_t) f(X_t)\} \qquad x \in R^d,$$

is a classical Schrödinger semigroup and hence has a jointly continuous symmetric integral kernel $q^{(n)}(t, x, y)$ satisfying (0.1). We are going to show that the sequence $q^{(n)}$ converges uniformly on any compact subset of $(0, \infty) \times R^d \times R^d$ to a function q and this function q is an integral kernel of $(T_t)_{t>0}$. In order to make this strategy work, we have to show that the sequence of functions u_n we construct converges nicely to u (see Lemma (1.5) and Lemma (1.8)) and has some kind of uniformity (see Lemma (1.6)).

The organization of this paper is as follows. Section 1 serves as a preparation, some basic results on admissible functions are given there and these results will be essential in studying the generalzed Schrödinger semigroup. Section 2 contains the main results. Section 3 is an application of the results in Section 2, an improved version of the gauge theorem in [10] is proved there and some useful consequences of the gauge theorem are also given there.

1. PRELIMINARIES

In this paper we shall always assume that $X = (\Omega, \mathcal{F}, \mathcal{F}_t, X_t, \theta_t, P^x)$ is a standard Brownian motion in R^d, $d \geq 3$, and that $(\mathcal{E}, H^1(R^d))$ is the regular Dirichlet form associated with X. The Sobolev space $H^1(R^d)$ is defined by

$$H^1(R^d) = \{u \in L^2(R^d) : \frac{\partial u}{\partial x_i} \in L^2(R^d), i = 1, 2, \cdots, d\}$$

and $\mathcal{E}$ is defined by

$$\mathcal{E}(u, v) = \int_{R^d} \nabla u \cdot \nabla v dx, \qquad \forall u, v \in H^1(R^d).$$

For a positive number α, we shall write

$$\mathcal{E}_\alpha(u,v) = \mathcal{E}(u,v) + \alpha(u,v),$$

where (u,v) denotes the inner product in $L^2(R^d)$.

As usual we use $p(t,x,y)$ to denote the transition density of X_t. We shall sometimes write

$$P_t\mu(x) = \int_{R^d} p(t,x,y)\mu(dy)$$

and

$$G_\alpha\mu(x) = \int_0^\infty e^{-\alpha t} P_t\mu(x)dt$$

provided the right hand sides make sense.

If $u \in H^1(R^d)$ is bounded and continuous, then

$$A_t^{[u]} = u(X_t) - u(X_0) - \int_0^t \nabla u(X_s) \cdot dX_s$$

is a continuous additive functional of zero energy with respect to X. In general $A_t^{[u]}$ is not of bounded variation. We shall call $(A_t^{[u]})$ the continuous additive functional of zero energy generated by u.

(1.1) Definition. A signed Radon measure μ on R^d is said to be in the Kato class if

$$\limsup_{r \downarrow 0} \sup_{x \in R^d} \int_{|x-y|<r} \frac{\mu^*(dy)}{|x-y|^{d-2}} = 0$$

where $\mu^* = \mu^+ + \mu^-$ with μ^+ and μ^- being the positive part and the negative part of μ respectively. In the sequel we shall use K_d to denote the Kato class.

The following result is well known. See, for example, [3].

(1.2) Proposition. *Let μ be a Radon measure on R^d. Then $\mu \in K_d$ if and only if*

$$\lim_{t \downarrow 0} \sup_{x \in R^d} \int_0^t \int_{R^d} p(s,x,y)\mu^*(dy)ds = 0.$$

(1.3) Definition. A bounded continuous function $u \in H^1(R^d)$ is said to be admissible if $|\nabla u|^2(x)dx$ is in the Kato class K_d.

From now on we are going to fix an admissible function u.

Put

$$J(x) = \begin{cases} k\exp(\frac{-1}{1-|x|^2}) & |x| < 1; \\ 0, & |x| \geq 1 \end{cases}$$

where k is a positive constant such that

$$\int_{R^d} J(x)dx = 1.$$

For any positive integer n, put

$$J_n(x) = n^d J(nx),$$

then

(1) $J_n(x) = 0$ when $|x| \geq \frac{1}{n}$;
(2) $\int_{R^d} J_n(x)dx = 1$.

For any $n \geq 1$, define

$$I_n(x) = \begin{cases} 1, & |x| < n; \\ \exp(\frac{-1}{1-(|x|-n)^2}), & n \leq |x| < n+1; \\ 0, & |x| \geq n+1. \end{cases}$$

Then it is easy to see that for any $1 \leq i \leq d$,

$$\frac{\partial I_n(x)}{\partial x_i} = \begin{cases} \frac{-2(|x|-n)x_i}{|x|(1-(|x|-n)^2)^2} \exp(\frac{-1}{1-(|x|-n)^2}), & n \leq |x| < n+1; \\ 0, & \text{otherwise.} \end{cases}$$

From this it easy to see that there exists a constant M such that for any $n \geq 1$ and any $1 \leq i \leq d$,

$$\|\frac{\partial I_n}{\partial x_i}\|_\infty \leq M^{\frac{1}{2}}. \tag{1.4}$$

Now put

$$v_n(x) = I_n(x) \cdot u(x),$$
$$u_n(x) = J_n * v_n(x).$$

Then the following result is classical. (See, for instance, [1].)

(1.5) Lemma.

(1) *For any n,*

$$\|u_n\|_\infty \leq \|u\|_\infty;$$

(2) *For any n, $u_n \in C_0^\infty(R^d)$, where $C_0^\infty(R^d)$ denotes the class of infinitely differentiable functions with compact support;*
(3) *$u_n \longrightarrow u$ uniformly on any compact subset of R^d;*
(4) *$u_n \longrightarrow u$ strongly in $H^1(R^d)$.*

Proof. Omitted.

(1.6) Lemma. *For the u_n's defined above, we have*

$$\limsup_{t\downarrow 0}\sup_{n,x} E^x \int_0^t |\nabla u_n|^2(X_s)ds = 0.$$

Proof. For any x and any n,

$$\begin{aligned}
&E^x \int_0^t |\nabla u_n|^2(X_s)ds \\
=& \int_0^t \int_{R^d} p(s,x,y)|\nabla u_n|^2(y)dyds \\
\leq& \int_0^T \int_{R^d} p(s,x,y) \int_{R^d} J_n(z)|\nabla v_n|^2(y-z)dzdyds \\
\leq& 2\int_0^t \int_{R^d} p(s,x,y) \int_{R^d} J_n(z)|\nabla u|^2(y-z)dzdyds \\
&+ 2dM \int_0^t \int_{R^d} p(s,x,y) \int_{R^d} J_n(z)u^2(y-z)dzdyds \\
=& 2C_1(x,t) + 2dMC_2(x,t),
\end{aligned}$$

where in the fourth line above we used (1.4) and the identity $\nabla v_n = u\nabla I_n + I_n \nabla u$.

Since

$$\begin{aligned}
C_1(x,t) &= \int_{R^d} J_n(z) \left(\int_0^t \int_{R^d} p(s,x,y)|\nabla u|^2(y-z)dyds \right) dz \\
&= \int_{R^d} J_n(z) \left(\int_0^t \int_{R^d} p(s,x,y+z)|\nabla u|^2(y)dyds \right) dz \\
&= \int_{R^d} J_n(z) \left(\int_0^t \int_{R^d} p(s,x-z,y)|\nabla u|^2(y)dyds \right) dz \\
&\leq \sup_x \int_0^t \int_{R^d} p(s,x,y)|\nabla u|^2(y)dyds,
\end{aligned}$$

and

$$\begin{aligned}
C_2(x,t) &= \int_{R^d} J_n(z) \left(\int_0^t \int_{R^d} p(s,x,y)u^2(y-z)dyds \right) dz \\
&= \int_{R^d} J_n(z) \left(\int_0^t \int_{R^d} p(s,x,y+z)u^2(y)dyds \right) dz \\
&= \int_{R^d} J_n(z) \left(\int_0^t \int_{R^d} p(s,x-z,y)u^2(y)dyds \right) dz \\
&\leq \sup_x \int_0^t \int_{R^d} p(s,x,y)u^2(y)dyds,
\end{aligned}$$

the assertion of this lemma follows from Proposition (1.2).

(1.7) Lemma. *For the u_n's defined above, we have*

$$\limsup_{\delta\downarrow 0}\sup_{n,x}\int_{|x-y|\le\delta}\frac{|\nabla u_n|^2(y)}{|x-y|^{d-2}}dy=0.$$

Proof. Follows easily from the previous lemma.

(1.8) Lemma. *We have*

$$\lim_{n\to\infty}\sup_x\int_{R^d}\frac{|\nabla(u_n-u)|^2(y)}{|x-y|^{d-2}}dy=0.$$

Proof. For any x,

$$\begin{aligned}&\int_{R^d}\frac{|\nabla(u_n-u)|^2(y)}{|x-y|^{d-2}}dy\\&=\int_{|x-y|\le\delta}\frac{|\nabla(u_n-u)|^2(y)}{|x-y|^{d-2}}dy+\int_{|x-y|>\delta}\frac{|\nabla(u_n-u)|^2(y)}{|x-y|^{d-2}}dy\\&=B_1(\delta)+B_2(\delta).\end{aligned}$$

It is easy to see that

$$B_1(\delta)\le 2\int_{|x-y|\le\delta}\frac{|\nabla u_n|^2(y)+|\nabla u|^2(y)}{|x-y|^{d-2}}dy$$

and

$$\begin{aligned}B_2(\delta)&\le\delta^{-d+2}\int_{R^d}|\nabla(u_n-u)|^2(y)dy\\&=\delta^{-d+2}\int_{|x-y|>r}|\nabla(u_n-u)|^2(y)dy+\delta^{-d+2}\int_{|x-y|\le r}|\nabla(u_n-u)|^2(y)dy\\&=\delta^{-d+2}B_3(r)+\delta^{-d+2}B_4(r).\end{aligned}$$

Now

$$\begin{aligned}B_3(r)&\le 2\int_{|x-y|>r}(|\nabla u_n|^2(y)+|\nabla u|^2(y))dy\\&\le 2\int_{|x-y|>r}|\nabla u|^2(y)dy+2\int_{|x-y|>r}\int_{R^d}J_n(z)|\nabla v_n(y-z)|^2dzdy\\&=B_5(r)+B_6(r),\end{aligned}$$

and

$$B_6(r) \le 4\int_{|x-y|>r}\int_{R^d} J_n(z)|\nabla u|^2(y-z)dzdy+$$
$$+ 4dM\int_{|x-y|>r}\int_{R^d} J_n(z)u^2(y-z)dzdy$$
$$= 4\int_{|z|<\frac{1}{n}} J_n(z)\int_{|x-y|>r} |\nabla u|^2(y-z)dzdy+$$
$$+ 4dM\int_{|z|<\frac{1}{n}} J_n(z)\int_{|x-y|>r} u^2(y-z)dzdy$$
$$\le 4\int_{|y|>r-1} |\nabla u|^2(y-z)dzdy + 4dM\int_{|y|>r-1} u^2(y)dy,$$

where in the second line we used (1.4) and the identity $\nabla v_n = u\nabla I_n + I_n\nabla u$.
For any $\epsilon > 0$, we can take, by Lemma (1.7), a $\delta > 0$ such that

$$\sup_{n,x}\int_{|x-y|\le\delta} \frac{\nabla u_n|^2(y) + |\nabla u|^2(y)}{|x-y|^{d-2}}dy < \frac{\epsilon}{8},$$

consequently

$$B_1(\delta) < \frac{\epsilon}{4}.$$

Now we fix this δ. Since $u \in H^1(R^d)$, we know that there exists an r such that

$$\int_{|y|>r-1} |\nabla u|^2(y)dy \le \frac{\epsilon}{24}\delta^{d-2},$$
$$\int_{|y|>r-1} u^2(y) \le \frac{\epsilon}{16dM}\delta^{d-2},$$

consequently

$$B_3(r) \le \frac{\epsilon}{2}\delta^{d-2}.$$

Now we fix this r. Since $u_n \longrightarrow u$ in $H^1(B(0,r))$, we know that there exists an N such that

$$\int_{|y|\le r} |\nabla(u_n - u)|^2(y)dy \le \frac{\epsilon}{4}\delta^{d-2},$$

whenever $n > N$. Therefore

$$\sup_x \int_{R^d} \frac{|\nabla(u_n-u)|^2(y)}{|x-y|^{d-2}}dy \le \epsilon,$$

whenever $n > N$. The proof is now complete.

2. Schrödinger semigroups

A continuous additive functional (A_t) is said to be a special additive functional if it can be decomposed as the sum of two additive functionals $(\tilde{A}_t)$ and $(\overline{A}_t)$, with $(\tilde{A}_t)$ being generated by an admissible function and $(\overline{A}_t)$ being associated with a measure in the Kato class. In this paper we are doing to fix an special additive functional (A_t) and study the following generalized Schrödinger semigroup:

$$T_t f(x) = E^x\{e^{A_t} f(X_t)\}.$$

The idea of this paper is to approximate A_t by a sequence of additive functionals associated with measures in the Kato class. Fix a decomposition of (A_t):

$$A_t = \tilde{A}_t + \overline{A}_t$$

with $\tilde{A}_t = A_t^{[u]}$ for some admissible u and $\overline{A}_t$ being associated with $\mu \in K_d$. Define u_n as in Section 1 and put, for $n \geq 1$,

$$\tilde{A}_t^{(n)} = A_t^{[u_n]},$$
$$A_t^{(n)} = \tilde{A}_t^{(n)} + \overline{A}_t,$$
$$T_t^{(n)} f(x) = E^x\{e^{A_t^{(n)}} f(X_t)\}.$$

(2.1) Lemma. *For each $n \geq 1$, $T_t^{(n)}$ possesses an integral kernel $q^{(n)}(t, x, y)$ such that*

(1) $q^{(n)}$ is continuous on $(0, \infty) \times R^d \times R^d$;

(2) for any $t > 0$ and any $x, y \in R^d$,

$$q^{(n)}(t, x, y) = q^{(n)}(t, y, x).$$

Furthermore, there exist constants C and β independent of n such that for any $n \geq 1$ and any $x, y \in R^d$ we have

$$q^{(n)}(t, x, y) \leq Ce^{\beta t} t^{-\frac{d}{2}}$$

Proof. The first part is well known (see, for instance, [3]), so we need only prove the second part.

It follows from Lemma (1.6) that there exists a $t_0 > 0$ such that for any $t \leq t_0$,

$$\sup_{n,x} E^x 16 \int_0^t |\nabla u_n(X_s)|^2 ds \leq c_1^2 < \frac{1}{16},$$

consequently, for any $t \leq t_0$,

$$\sup_{n,x} E^x \left\{4 \left| \int_0^t \nabla u(X_s) \cdot dX_s \right| \right\} \leq c_1 < \frac{1}{4}.$$

Now for any stopping time $S \le T \le t_0$,

$$\begin{aligned}
&E^x\left\{\left(4\left|\int_S^T \nabla u_n(X_s)\cdot dX_s\right|\right)|\mathcal{F}_S\right\}\\
&\le \left(E^x\left\{(16\int_S^T \nabla u_n(X_s)\cdot dX_s)^2|\mathcal{F}_S\right\}\right)^{\frac{1}{2}}\\
&= \left(E^x\left\{16\int_S^T |\nabla u_n(X_s)|^2 ds|\mathcal{F}_S\right\}\right)^{\frac{1}{2}}\\
&\le \left(E^x\left\{16\int_S^{t_0+S} |\nabla u_n(X_s)|^2 ds|\mathcal{F}_S\right\}\right)^{\frac{1}{2}}\\
&= \left(E^{X_S}16\int_0^{t_0} |\nabla u_n(X_s)|^2 ds\right)^{\frac{1}{2}}\\
&\le c_1 < \frac{1}{4}.
\end{aligned}$$

Therefore by the John–Nirenberg inequality (see [8]) we know that

$$\sup_{n,x} E^x\left\{\exp(\sup_{t\le t_0} 4\left|\int_0^t \nabla u_n(X_s)\cdot dX_s\right|)\right\} \le \frac{1}{1-4c_1} := c_2.$$

In particular

$$\sup_{n,x} E^x\left\{\sup_{t\le t_0}\exp(-4\int_0^t \nabla u_n(X_s)\cdot dX_s)\right\} \le c_2.$$

From [3] we know that

$$\sup_x E^x\{\sup_{t\le t_0} e^{4\overline{A}_t}\} \le c_3.$$

Thus

$$\begin{aligned}
&\sup_{n,x} E^x\left\{\sup_{t\le t_0} e^{2A_t^{(n)}}\right\}\\
&\le \left(\sup_x E^x\{\sup_{t\le t_0} e^{4\overline{A}_t}\}\right)^{\frac{1}{2}} \sup_{n,x}\left(E^x\{\sup_{t\le t_0} e^{4\tilde{A}_t^{(n)}}\}\right)^{\frac{1}{2}}\\
&\le c_3^{\frac{1}{2}} e^{4\|u_n\|_\infty} \sup_{n,x}\left(E^x\{\sup_{t\le t_0}\exp(-4\int_0^t \nabla u(X_s)\cdot dX_s)\}\right)^{\frac{1}{2}}\\
&\le c_3^{\frac{1}{2}} c_2^{\frac{1}{2}} e^{4\|u_n\|_\infty} := c_4.
\end{aligned}$$

From the above we immediately get

$$\sup_{n,x} E^x \left\{ \sup_{t \le t_0} e^{A_t^{(n)}} \right\} \le c_4^{\frac{1}{2}} := c_5. \tag{2.2}$$

For any $f \in L^2(R^d)$ and $t \le t_0$,

$$\begin{aligned}
\sup_{n,x} |T_t^{(n)} f(x)| &\le \sup_{n,x} E^x \left\{ e^{A_t^{(n)}} |f|(X_t) \right\} \\
&\le \sup_{n,x} \left(E^x \left\{ e^{2A_t^{(n)}} \right\} \right)^{\frac{1}{2}} \left(E^x f^2(X_t) \right)^{\frac{1}{2}} \\
&\le c_4^{\frac{1}{2}} \frac{1}{(2\pi t)^{\frac{d}{4}}} \|f\|_2 := c_6 t^{-\frac{d}{4}} \|f\|_2 .
\end{aligned}$$

Therefore for any $t \le t_0$,

$$\|T_t^{(n)}\|_{2,\infty} \le c_6 t^{-\frac{d}{4}}.$$

From this and the self-adjointness of $T_t^{(n)}$ we get that for any $t \le t_0$,

$$\|T_t^{(n)}\|_{1,2} = \|T_t^{(n)}\|_{2,\infty} \le c_6 t^{-\frac{d}{4}}.$$

Therefore by the semigroup property we have that for any $t \le t_0$,

$$\begin{aligned}
\|T_t^{(n)}\|_{1,\infty} &\le \|T_{\frac{t}{2}}^{(n)}\|_{1,2} \|T_{\frac{t}{2}}^{(n)}\|_{2,\infty} \\
&\le c_6^2 (\frac{t}{2})^{-\frac{d}{2}} := c_7 t^{-\frac{d}{2}}.
\end{aligned}$$

Now using the semigroup property again we can get that for any $k \ge 1$ and $t \le t_0$,

$$\|T_{kt_0+t}^{(n)}\|_{1,\infty} \le c_7 t_0^{-\frac{d}{2}} (c_5)^k.$$

Setting

$$\begin{aligned}
C &= c_7 t_0^{-\frac{d}{2}}, \\
\beta &= \frac{d}{2} + t_0^{-1} \ln c_5,
\end{aligned}$$

we get that for any $t > 0$

$$\|T_t^{(n)}\|_{1,\infty} \le C e^{\beta t} t^{-\frac{d}{2}}. \tag{2.3}$$

Hence it follows from the Dunford and Pettis theorem (see, for instance, [13]) that for any n and any x, y,

$$q^{(n)}(t, x, y) \le C e^{\beta t} t^{-\frac{d}{2}}.$$

From (2.2) we can easily prove the following result.

(2.4) Lemma. *There exist C and β such that*

$$\sup_{n,x} E^x e^{A_t^{(n)}} \leq Ce^{\beta t}.$$

Proof. Omitted.

(2.5) Theorem. *For any $t > 0$, we have*

$$\sup_x E^x \left\{ \sup_{s \leq t} e^{A_s} \right\} < \infty.$$

Proof. As in the first part of the proof of Theorem (2.1) we can show that there exists a $t_0 > 0$ such that

$$\sup_x E^x \left\{ \sup_{s \leq t_0} e^{A_s} \right\} := M < \infty.$$

From the above we immediately get that there exist C and β such that for any $t > 0$

$$\sup_x E^x e^{A_t} \leq Ce^{\beta t}.$$

Now for any $t > 0$, there exists an integer n such that $t \leq nt_0$, where t_0 is specified above. Then

$$\begin{aligned}
\sup_{x \in R^d} E^x \left\{ \sup_{s \leq t} e(s) \right\} &\leq \sup_{x \in R^d} E^x \left\{ \sup_{s \leq nt_0} e(s) \right\} \\
&\leq \sum_{k=0}^{n-1} \sup_{x \in R^d} E^x \left\{ \sup_{kt_0 < s \leq (k+1)t_0} e(s) \right\} \\
&= \sum_{k=0}^{n-1} \sup_{x \in R^d} E^x \left\{ e(kt_0) E^{X(kt_0)} (\sup_{s \leq t_0} e(s)) \right\} \\
&\leq M \sum_{k=0}^{n-1} \sup_{x \in R^d} E^x \{ e(kt_0) \} \\
&\leq M \sum_{k=0}^{n-1} Ce^{k\beta t_0} < \infty.
\end{aligned}$$

The following result will be used to prove that $q^{(n)}$ converges uniformly on any compact subset of $(0, \infty) \times R^d \times R^d$.

(2.6) Lemma. *Suppose $T > 0$ is fixed and K is a relatively compact, open subset of R^d. Then for any $\epsilon > 0$, there exists an N such that*

$$\sup_{t\leq T} \|T_t^{(n)} - T_t^{(m)}\|_{L^\infty(K),L^\infty(K)} < \epsilon$$

whenever $n, m > N$.

Proof. For any $f \in L^\infty(R^d)$, any $t \leq T$ and any relatively compact, open subset G having the closure of K as a subset,

$$\begin{aligned}
&\sup_{x\in K} E^x\left\{(e^{A_t^{(n)}} - e^{A_t^{(m)}})f(X_t)1_K(X_t)\right\}\\
&\leq \|f\|_{L^\infty(K)} \sup_{x\in K} E^x\left\{e^{A_t^{(n)}}|1 - e^{\tilde{A}_t^{(m)}-\tilde{A}_t^{(n)}}|1_G(X_t)\right\}\\
&\leq \|f\|_{L^\infty(K)} \sup_{x\in K}\left(E^x e^{2A_t^{(n)}}\right)^{\frac12} \sup_{x\in K}\left(E^x\left\{(1 - e^{\tilde{A}_t^{(m)}-\tilde{A}_t^{(n)}})^2 1_G(X_t)\right\}\right)^{\frac12}\\
&\leq C\|f\|_{L^\infty(K)} \sup_{x\in K}\left(E^x\left\{(e^{|\tilde{A}_t^{(m)}-\tilde{A}_t^{(n)}|} - 1)^2 1_G(X_t)\right\}\right)^{\frac12}\\
&\leq C\|f\|_{L^\infty(K)} \sup_{x\in K}\left(E^x\left\{e^{2|\tilde{A}_t^{(m)}-\tilde{A}_t^{(n)}|}1_G(X_t)\right\} - E^x 1_G(X_t)\right)^{\frac12}\\
&\leq C\|f\|_{L^\infty(K)}\left(\sup_{x\in K} E^x\left\{e^{2|\tilde{A}_t^{(m)}-\tilde{A}_t^{(n)}|}1_G(X_t)\right\} - \inf_{x\in K} P^x(X_t\in G)\right)^{\frac12}\\
&\leq C\|f\|_{L^\infty(K)}\left(\sup_{x\in K} E^x\left\{e^{2|\tilde{A}_t^{(m)}-\tilde{A}_t^{(n)}|}1_G(X_t) - 1\right\} + \right.\\
&\quad\left. + \left\{1 - \inf_{x\in K} P^x(X_t\in G)\right\}\right)^{\frac12},
\end{aligned}$$

where in the fourth line we used Lemma (2.4).

Take G so that

$$1 - \inf_{x\in K} P^x(X_t\in G) \leq \frac{\epsilon^2}{2C^2},$$

and then fix this G. Since

$$\lim_{\eta\downarrow 0}\frac{e^{4\eta}}{1-4\eta} = 1,$$

we can take an η with $0 < \eta < \frac14$ such that

$$\frac{e^{4\eta}}{1-4\eta} < 1 + \frac{\epsilon^2}{2C^2}.$$

Now since $u_n \longrightarrow u$ uniformly on G, we know that there exists an N_1 such that

$$\sup_{x\in G} |u_n(x) - u_m(x)| < \eta,$$

whenever $n, m > N_1$. By Lemma (1.8) we know that there exists an N_2 such that

$$\sup_x E^x \int_0^T 4|\nabla(u_m - u_n)|^2(X_s)ds < \eta^2,$$

whenever $n, m > N_2$. Repeating part of the argument in the proof of Lemma (2.1), we get

$$\sup_x E^x \left\{ \sup_{t\le T} e^{2|\int_0^t \nabla(u_m-u_n)(X_s)\cdot dX_s|} \right\} < \frac{1}{1-4\eta},$$

whenever $n, m > N_2$. Take $N = N_1 \vee N_2$. Then for any $t \le T$,

$$\begin{aligned}
&\sup_x E^x \left\{ e^{2|\tilde{A}_t^{(m)} - \tilde{A}_t^{(n)}|} 1_G(X_t) \right\} \\
&\le e^{4\eta} \sup_x E^x \left\{ e^{2|\int_0^t \nabla(u_m-u_n)(X_s)\cdot dX_s|} \right\} \\
&\le \frac{e^{4\eta}}{1-4\eta} < 1 + \frac{\epsilon^2}{2C^2},
\end{aligned}$$

whenever $n, m > N$. Therefore for any $t \le T$ and any $f \in L^\infty(R^d)$ we have

$$\sup_{x\in K} E^x \left\{ (e^{A_t^{(n)}} - e^{A_t^{(m)}}) f(X_t) 1_K(X_t) \right\} \le \epsilon \|f\|_{L^\infty(K)},$$

whenever $n, m > N$.

(2.7) Lemma. *Suppose $0 < T_1 < T_2$ are fixed and K is a relatively compact, open subset of R^d. Then for any $\epsilon > 0$, there exists an N such that*

$$\sup_{T_1\le t\le T_2} \|T_t^{(n)} - T_t^{(m)}\|_{L^1(K),L^\infty(K)} < \epsilon,$$

whenever $n, m > N$; or equivalently

$$\sup_{T_1\le t\le T_2} \sup_{x,y\in K} |q^{(n)}(t,x,y) - q^{(m)}(t,x,y)| < \epsilon,$$

whenever $n, m > N$.

Proof. For any $t \in [T_1, T_2]$, we have

$$\begin{aligned}
&\|T_t^{(n)} - T_t^{(m)}\|_{L^1(K),L^\infty(K)} \\
&\le \|T_{\frac{t}{2}}^{(n)}(T_{\frac{t}{2}}^{(n)} - T_{\frac{t}{2}}^{(m)})\|_{L^1(K),L^\infty(K)} + \|(T_{\frac{t}{2}}^{(n)} - T_{\frac{t}{2}}^{(m)})T_{\frac{t}{2}}^{(m)}\|_{L^1(K),L^\infty(K)} \\
&\le \|T_{\frac{t}{2}}^{(n)}\|_{1,\infty} \|T_{\frac{t}{2}}^{(n)} - T_{\frac{t}{2}}^{(m)}\|_{L^1(K),L^1(K)} + \\
&\quad + \|T_{\frac{t}{2}}^{(n)} - T_{\frac{t}{2}}^{(m)}\|_{L^\infty(K),L^\infty(K)} \|T_{\frac{t}{2}}^{(m)}\|_{1,\infty}.
\end{aligned}$$

But by duality

$$\|T^{(n)}_{\frac{t}{2}} - T^{(m)}_{\frac{t}{2}}\|_{L^1(K),L^1(K)} = \|T^{(n)}_{\frac{t}{2}} - T^{(m)}_{\frac{t}{2}}\|_{L^\infty(K),L^\infty(K)}.$$

Therefore by (2.3) we know that for any $t \in [T_1, T_2]$,

$$\begin{aligned}
&\|T^{(n)}_t - T^{(m)}_t\|_{L^1(K),L^\infty(K)} \\
&\le (\|T^{(n)}_{\frac{t}{2}}\|_{1,\infty} + \|T^{(m)}_{\frac{t}{2}}\|_{1,\infty})(\|T^{(n)}_{\frac{t}{2}} - T^{(m)}_{\frac{t}{2}}\|_{L^\infty(K),L^\infty(K)}) \\
&\le Ce^{\beta T_2}(\frac{T_1}{2})^{-\frac{d}{2}}(\|T^{(n)}_{\frac{t}{2}} - T^{(m)}_{\frac{t}{2}}\|_{L^\infty(K),L^\infty(K)}).
\end{aligned}$$

Now the assertion of this Lemma follows immediately from Lemma (2.6).

(2.8) Theorem. *For each $t > 0$, T_t possesses an integral kernel $q(t, x, y)$ such that*

(1) *q is jointly continuous on $(0, \infty) \times R^d \times R^d$;*

(2) *there exist C and β such that for all $t > 0$ and all $x, y \in R^d$,*

$$q(t, x, y) \le Ce^{\beta t}t^{-\frac{d}{2}};$$

(3) *q is symmetric in x and y.*

Proof. Put

$$q(t, x, y) = \lim_{n\to\infty} q^{(n)}(t, x, y).$$

Then it follows from Lemma (2.1) and Lemma (2.7) that q satisfies the three conditions in the theorem. Now we are going to show that for any $t > 0$, $q(t, \cdot, \cdot)$ is indeed the integral kernel of T_t. By applying Lemma (2.4) to $2A^{(n)}_t$ we get that for any $t > 0$ and any x, $\{e^{A^{(n)}_t}; n \ge 1\}$ is uniformly integrable with respect to P^x. Thus for any $t > 0$, any x and any $f \in L^\infty(R^d)$,

$$\begin{aligned}
T_t f(x) &= E^x\left\{e^{A_t} f(X_t)\right\} \\
&= \lim_{n\to\infty} E^x\left\{e^{A^{(n)}_t} f(X_t)\right\}.
\end{aligned}$$

Therefore by the dominated convergence theorem we know that for any $t > 0$, any x and any bounded, nonnegative function $f \in L^1(R^d)$,

$$\begin{aligned}
T_t f(x) &= \lim_{n\to\infty} \int_{R^d} q^{(n)}(t, x, y) f(y) dy \\
&= \int_{R^d} q(t, x, y) f(y) dy.
\end{aligned}$$

Using the monotone convergence theorem we can get that for any $t > 0$, any x and any nonnegative Borel function f,

$$T_t f(x) = \int_{R^d} q(t, x, y) f(y) dy,$$

i.e., $q(t, \cdot, \cdot)$ is the integral kernel of T_t for any $t > 0$.

From the above lemma it is easy to see that q is strictly positive on $(0, \infty) \times R^d \times R^d$. The following result claims even more.

(2.9) Theorem. *There exist C and β such that for all $t > 0$ and all $x, y \in R^d$,*

$$C^{-1}e^{-\beta t}t^{-\frac{d}{2}}\exp(\frac{-3|x-y|^2}{4t}) \leq q(t,x,y) \leq Ce^{\beta t}t^{-\frac{d}{2}}\exp(\frac{-|x-y|^2}{4t}).$$

Proof. Fix $x, y \in R^d$ and put

$$B_r = \{z \in R^d : |z-y| \leq r\}$$
$$|B_r| = \int_{B_r} dz.$$

Then

$$\begin{aligned}
&\frac{1}{|B_r|}\int_{B_r} q(t,x,z)dz \\
&= \frac{1}{|B_r|}E^x\left\{e^{A_t}1_{B_r}(X_t)\right\} \\
&\leq \left(\frac{1}{|B_r|}E^x\left\{e^{2A_t}1_{B_r}(X_t)\right\}\right)^{\frac{1}{2}}\left(\frac{1}{|B_r|}E^x 1_{B_r}(X_t)\right)^{\frac{1}{2}} \\
&\leq \left(\frac{1}{|B_r|}\int_{B_r} C_1 e^{\beta_1 t}t^{-\frac{d}{2}}dz\right)^{\frac{1}{2}}\left(\frac{1}{|B_r|}\int_{B_r}(2\pi t)^{-\frac{d}{2}}\exp(\frac{-|x-z|^2}{2t})dz\right)^{\frac{1}{2}}.
\end{aligned}$$

Letting $r \downarrow 0$ we get the second inequality of the theorem. Similarly we have

$$\begin{aligned}
&\frac{1}{|B_r|}E^x 1_{B_r}(X_t) \\
&\leq \left(\frac{1}{|B_r|}E^x e^{-A_t}1_{B_r}(X_t)\right)^{\frac{1}{2}}\left(\frac{1}{|B_r|}E^x e^{A_t}1_{B_r}(X_t)\right)^{\frac{1}{2}},
\end{aligned}$$

consequently

$$\begin{aligned}
&\frac{1}{|B_r|}\int_{B_r} q(t,x,z)dz \\
&\geq \frac{\left(\frac{1}{|B_r|}\int_{B_r}(2\pi t)^{-\frac{d}{2}}\exp(\frac{-|x-z|^2}{2t})dz\right)^2}{\frac{1}{|B_r|}\int_{B_r} C_2 e^{\beta_2 t}t^{-\frac{d}{2}}\exp(\frac{-|x-z|^2}{4t})dz}.
\end{aligned}$$

Letting $r \downarrow 0$, we get the first inequality of the theorem

The following results are important consequences of of Theorems (2.8) and (2.9).

(2.10) Theorem.

(1) *For each* $1 \leq p \leq p' \leq \infty$, T_t *is a bounded operator from* $L^p(R^d)$ *to* $L^{p'}(R^d)$;

(2) T_t *sends* $L^p(R^d)$ *to* $C_\infty(R^d)$ *for all* $1 \leq p < \infty$ *and sends* $L^\infty(R^d)$ *to* $C_b(R^d)$. *Here* $C_\infty(R^d)$ *is the family of all continuous functions* f *such that* $\lim_{|x|\to\infty} f(x) = 0$, *and* $C_b(R^d)$ *is the family of all bounded continuous functions on* R^d;

(3) $(T_t)_{t>0}$ *is a strongly continuous semigroup on* $L^p(R^d)$ *for each* $1 \leq p < \infty$, *a strongly continuous semigroup on* $C_\infty(R^d)$ *and a weakly continuous semigroup on* $L^\infty(R^d)$;

(4) T_t *is a self-adjoint operator on* $L^2(R^d)$.

Proof. By Applying Theorems (2.8) and (2.9).

Let D be a connected open subset of R^d and let τ_D be the first time X exits from D:

$$\tau_D = \inf\{t > 0 : X_t \notin D\}.$$

Similar to $(T_t)_{t>0}$ we can study the following semigroup:

$$T_t^D f(x) = E^x\left\{e^{A_t} f(X_t); t < \tau_D\right\}, \qquad x \in D.$$

In fact,if we define

$$T_t^{(D,n)} f(x) = E^x\{e^{A_t^{(n)}} f(X_t); t < \tau_D\}, \qquad x \in D$$

and if we use $q^{(D,n)}(t,x,y)$ to denote the integral kernel of $T_t^{(D,n)}$, then we can get the following results by using the same arguments.

(2.11) Theorem. *Suppose* $0 < T_1 < T_2$ *are fixed and* K *is a relatively compact, open subset of* D. *Then for any* $\epsilon > 0$, *there exists an* N *such that*

$$\sup_{T_1 \leq t \leq T_2} \|T_t^{(D,n)} - T_t^{(D,m)}\|_{L^1(K),L^\infty(K)} < \epsilon,$$

whenever $n, m > N$; *or equivalently*

$$\sup_{T_1 \leq t \leq T_2} \sup_{x,y \in K} |q^{(D,n)}(t,x,y) - q^{(D,m)}(t,x,y)| < \epsilon,$$

whenever $n, m > N$.

(2.12) Theorem. *For each* $t > 0$, T_t^D *possesses an integral kernel* $q^D(t,x,y)$ *such that*

(1) q^D *is jointly continuous on* $(0,\infty) \times D \times D$;

(2) *there exist* C *and* β *such that for any* $x, y \in D$,

$$0 < q^D(t,x,y) \leq Ce^{\beta t} t^{-\frac{d}{2}};$$

(3) *for any* $x, y \in D$,

$$q^D(t,x,y) = q^D(t,y,x).$$

In the remainder of this paper, β will always stands for the constant specified above.

(2.13) Theorem.

(1) *For each* $1 \le p \le p' \le \infty$ *and any* $t > 0$, T_t^D *is a bounded operator from* $L^p(D)$ *to* $L^{p'}(D)$.
(2) *For each* $1 \le p \le \infty$ *and any* $t > 0$, T_t^D *sends* $L^p(D)$ *into* $C_b(D)$, *where* $C_b(D)$ *denotes the class of all bounded continuous functions on* D.
(3) *If* $x_0 \in \partial D$ *is a regular point, then for any* $1 \le p\infty$ *and any* $t > 0$,

$$\lim_{D \ni x \to x_0} |T_t^D f(x)| = 0.$$

(4) $(T_t^D)_{t>0}$ *is a strongly continuous semigroup on* $L^p(D)$ *for each* $1 \le p < \infty$.
(5) *For any* $t > 0$, T_t^D *is a self–adjoint operator on* $L^2(D)$.

3. The Gauge Theorem

In this section we are going to assume that D is a bounded connected open subset of R^d. Under this assumption, $(T_t^D)_{t>0}$ is a strongly continuous semigroup of compact self–adjoint operators. Denote by L the generator of $(T_t^D)_{t>0}$ on $L^2(D)$ and by $\mathcal{D}(L)$ the domain of L. Our first observation is that all the eigenvalues of L are real numbers.

For any $\alpha > 0$ and any Borel function f on D, we define

$$G_\alpha^D f(x) = \int_0^\infty e^{-\alpha t} T_t^D f(x) dt, \qquad x \in D,$$

provided the right hand side makes sense. Then the following is classical.

(3.1) Propostion. *Let* $\phi \in L^2(D)$ *and* λ *be a real number. Then the following statements are equivalent.*

(1) $T_t^D \phi = e^{\lambda t}\phi$ *for some* $t > 0$;
(2) $T_t^D \phi = e^{\lambda t}\phi$ *for all* $t > 0$;
(3) $\phi \in \mathcal{D}(L)$ *and* $L\phi = \lambda\phi$;
(4) $G_\alpha^D \phi = (\alpha - \lambda)^{-1}\phi$ *for some* $\alpha > (\lambda \vee \beta)$;
(5) $G_\alpha^D \phi = (\alpha - \lambda)^{-1}\phi$ *for all* $\alpha > (\lambda \vee \beta)$.

Recall that in this section β *is always specified by Theorem (2.12).*

(3.2) Lemma. *For all* $\alpha > \beta$, *all the eigenvalues of* G_α^D *are strictly positive.*

Proof. Let $\phi \in L^2(D)$. Then

$$\begin{aligned} &\int_D \phi(x) G_\alpha^D \phi(x) dx \\ &= \int_0^\infty e^{-\alpha t} \int_D \phi(x) T_t^D \phi(x) dx dt \\ &= \int_0^\infty e^{-\alpha t} \int_D (T_{\frac{t}{2}}^D \phi(x))^2 dx dt \ge 0, \end{aligned}$$

consequently all the eigenvalues of G_α^D are nonnegative. But for all $\phi \in L^2(D)$ we have $(L-\alpha)G_\alpha^D\phi = -\phi$, thus zero cannot be an eigenvalue of G_α^D and the assertion follows.

(3.3) Proposition.

(1) *The operators L, $(T_t^D)_{t>0}$, $(G_\alpha^D)_{\alpha>\beta}$ admit a common set of eigenfunctions.*

(2) *All the eigenfunctions are bounded continuous on D.*

(3) *All the eigenvalues of L lie in the half line $(-\infty, \beta)$. If λ is an eigenvalue of L, then $e^{\lambda t}$ is an eigenvalue of T_t^D and $(\alpha-\lambda)^{-1}$ is an eignvalue of G_α^D.*

(4) *Let $\{\lambda_i\}_{i\geq 1}$ be all the eigenvalues of L written in decreasing order, each repeated according its multiplicity. Then $\lambda_i \downarrow -\infty$ and the corresponding eigenfunctions $\{\phi_i\}_{i\geq 1}$ can be so chosen that they form an orthonormal basis of $L^2(D)$.*

(5) *The eigenspace corresponding to the first eigenvalue is one dimensional and ϕ_1 can be chosen to be nonnegative.*

Proof. Follows easily from Proposition (3.1) and Lemma (3.2).

The following result is an important consequence of Theorem (2.12).

(3.4) Theorem. *Suppose that ϕ is an eigenfunction of L corresponding to the eigenvalue λ. If ϕ is nonnegative, then ϕ is strictly positive on D.*

Proof. If $\phi(x) = 0$ for some $x \in D$, then we have for any $t > 0$

$$0 = e^{\lambda t}\phi(x) = T_t^D\phi(x) = \int_D q^D(t,x,y)\phi(y)dy.$$

But by Theorem (2.12) $q^D(t,x,\cdot)$ is strictly positive on D, consequently $\phi(y) = 0$ for all $y \in D$, which contradicts the fact that ϕ is an eigenfunction of L.

It follows from the two results above that ϕ_1 can be chosen to be strictly positive. In the sequel we shall always assume that $\{\lambda_i\}_{i\geq 1}$ and $\{\phi_i\}_{i\geq 1}$ are as in Proposition (3.3) with ϕ_1 being strictly positive on D.

Let $f \in L^2(D)$. Then there exist a_i, $i = 1, 2, \cdots$ such that for almost every $x \in D$,

$$f(x) = \sum_{i=1}^{\infty} a_i\phi_i(x).$$

By Propositions (3.1) and (3.3) we know that for every $x \in D$, any $t > 0$ and any $i \geq 1$

$$T_t^D\phi_i(x) = e^{\lambda_i t}\phi_i(x), \tag{3.5}$$

consequently, for any $t > 0$ and almost every $x \in D$,

$$T_t^D f(x) = \sum_{i=1}^{\infty} a_i e^{\lambda_i t}\phi_i(x).$$

The following is a strengthening of the above identity.

(3.6) Theorem. *If $f = \sum_{i=1}^{\infty} a_i\phi_i \in L^2(D)$, then for any $t > 0$ and any $x \in D$,*

$$T_t^D f(x) = \sum_{i=1}^{\infty} a_i e^{\lambda_i t}\phi_i(x). \tag{3.7}$$

Moreover, for each compact subset K of $(0, \infty)$, we have

$$\lim_{n\to\infty} \sup_{x\in D} |T_t^D f(x) - \sum_{i=1}^{n} a_i e^{\lambda_i t}\phi_i(x)| = 0. \tag{3.8}$$

Proof. (3.8) follows easily from (3.5) and Theorem (2.12). (3.7) is an immediate consequence of (3.8).

(3.9) Theorem. *For any $t > 0$ and any $x, y \in D$,*

$$q^D(t, x, y) = \sum_{i=1}^{\infty} a_i\phi_i(x)\phi_i(y).$$

Proof. For any $\epsilon > 0$ and any $x \in D$,

$$\int_D q^D(\epsilon, x, y)\phi_i(y)dy = e^{\lambda_i \epsilon}\phi_i(x),$$

consequently, $q(\epsilon, x, \cdot) \in L^2(D)$ has the following expansion:

$$q(\epsilon, x, \cdot) = \sum_{i=1}^{\infty} e^{\lambda_i \epsilon}\phi_i(x)\phi_(\cdot).$$

Applying Theorem (3.6) we get

$$\begin{aligned} q^D(t+\epsilon, x, y) &= T_t^D(q^D(\epsilon, x, \cdot))(y) \\ &= \sum_{i=1}^{\infty} e^{\lambda_i(t+\epsilon)}\phi_i(x)\phi_i(y). \end{aligned}$$

(3.10) Theorem.

(1) *Let $f \in L^2(D)$ and λ be a real number. Then for each $x \in D$, the limit*

$$S_\lambda f(x) := \lim_{t\uparrow\infty} e^{-\lambda t}T_t^D f(x) \tag{3.11}$$

always exists in $[-\infty, \infty]$. If $|S_\lambda f(x)| < \infty$ on a dense subset of D, then

$$\lim_{t\uparrow\infty} e^{-\lambda t}|T_t^D f(x)|$$

exists and is finite.

(2) *Let* $f \in L^2(D)$ *with* $\|f\|_{L^2} > 0$. *Define*

$$\gamma = \inf\{\lambda \in R : \lim_{t\uparrow\infty} e^{-\lambda t}|T_t^D f(x)| < \infty\}.$$

Then $S_\gamma f$ *is an eigenfunction of* L *corresponding to the eigenfunction* γ.

Proof. Let $f = \sum_{i=1}^\infty a_i\phi_i$. Without loss of generality we may assume that there exists at least one $j \geq 1$ such that $\sum_{\lambda_i=\lambda_j} a_i\phi_i(x) \neq 0$ for some $x \in D$. Otherwise we have $e^{-\lambda t}T_t^D f(x) = 0$ for all $t > 0$ and all $x \in D$, and thus the limit (3.11) exists trivially. Setting

$$k = \min\{j \geq 1 : \sum_{\lambda_i=\lambda_j} a_i\phi_i(x) \neq 0 \text{ for some } x \in D\}.$$

Then by Theorem (3.6) we have for $0 < \epsilon < t$ and any $x \in D$,

$$\begin{aligned} &e^{-\lambda t}T_t^D f(x) \\ (3.12)\quad &= e^{-\lambda t}\left(\sum_{i=1}^\infty e^{\lambda_i t}a_i\phi_i(x)\right) \\ &= e^{(\lambda_k-\lambda)(t-\epsilon)-\lambda\epsilon}\left(\sum_{\lambda_i=\lambda_k} e^{\lambda_i\epsilon}a_i\phi_i(x) + \sum_{\lambda_i<\lambda_k} e^{(\lambda_i-\lambda_k)(t-\epsilon)}e^{\lambda_i\epsilon}a_i\phi_i(x)\right). \end{aligned}$$

Let

$$\eta = \min\{\lambda_k - \lambda_i : \lambda_i < \lambda_k\}.$$

Then $\eta > 0$ Since for any $x \in D$,

$$\begin{aligned} &\left(\sum_{i=1}^\infty |e^{\lambda_i\epsilon}a_i\phi_i(x)|\right)^2 \\ &\leq \left(\sum_{i=1}^\infty a_i^2\right)\left(\sum_{i=1}^\infty |e^{\lambda_i\epsilon}a_i\phi_i(x)|^2\right) \\ &= \|f\|_{L(D)}^2\|q^D(\epsilon, x, \cdot)\|_{L^2(D)}^2 \\ &\leq \|f\|_{L(D)}^2 Ce^{2\beta\epsilon}\epsilon^{-d}, \end{aligned}$$

and

$$\sum_{\lambda_i<\lambda_k} e^{(\lambda_i-\lambda_k)(t-\epsilon)}|e^{\lambda_i\epsilon}a_i\phi_i(x)| \leq e^{-\eta(t-\epsilon)}\sup_{x\in D}\sum_{i=1}^\infty |e^{\lambda_i\epsilon}a_i\phi_i(x)|,$$

we know that

$$\lim_{t\uparrow\infty} \sum_{\lambda_i<\lambda_k} e^{(\lambda_i-\lambda_k)(t-\epsilon)}|e^{\lambda_i\epsilon}a_i\phi_i(x)| = 0. \tag{3.13}$$

From (3.12) and (3.13) we get that

(a) If $\lambda_k - \lambda > 0$, then for any $x \in D$,

$$\lim_{t\uparrow\infty} e^{-\lambda t}T_t^D f(x) = \begin{cases} +\infty, & \text{if} \sum_{\lambda_i=\lambda_k} a_i\phi_(x) > 0; \\ -\infty, & \text{if} \sum_{\lambda_i=\lambda_k} a_i\phi_(x) < 0. \end{cases}$$

(b) If $\lambda_k - \lambda = 0$, then for any $x \in D$ we have

$$\lim_{t\uparrow\infty} e^{-\lambda t}T_t^D f(x) \sum_{\lambda_i=\lambda_k} a_i\phi_(x).$$

(c) If $\lambda_k - \lambda < 0$, then for any $x \in D$, we have

$$\lim_{t\uparrow\infty} e^{-\lambda t}T_t^D f(x) = 0.$$

Thus (3.11) is proved. The rest of the assertions can also be derived from (3.12) and (3.13).

(3.14) Lemma. *If $\phi \in L^2(D)$ is strictly positive on D, then the following assertions are equivalent.*

(1) *There exists an $x \in D$ such that*

$$\lim_{n\uparrow\infty} T_n^D \phi(x) = 0.$$

(2) $\lambda_1 < 0$.

Proof. Suppose that $\phi = \sum_{i=1}^{\infty} a_i\phi_i$, then by Theorem (3.6) we know that for any $t > 0$ and any $x \in D$,

$$T_t^D \phi(x) = \sum_{i=1}^{\infty} e^{\lambda_i t}a_i\phi_i(x).$$

The first eigenfunction ϕ_1 is strictly positive on D, therefore if ϕ is strictly positive on D, then $a_1 = \int_D \phi_1(x)\phi(x)dx > 0$. Thus from the proof of Theorem (3.10) we know that the assertions (1) and (2) are equivalent.

The following is an improved version of the gauge theorem in [10].

(3.15) Theorem. *The following assertions are equivalent.*

(1) *There exists a strictly positive bounded function f on ∂D such that*

$$E^x\left\{e^{A_{\tau_D}} f(X(\tau_D))\right\} < \infty$$

for some $x \in D$.

(2) *For all bounded function f on ∂D, the function*

$$x \mapsto E^x\left\{e^{A_{\tau_D}} f(X(\tau_D))\right\}$$

is bounded on D.

(3) *There exists a nonvanishing, nonnegative measure $\mu \in K_d$ such that*

$$E^x \int_0^{\tau_D} e^{A_s} dB_s < \infty$$

for some $x \in D$, where B_t is the positive continuous additive functional associated with μ.

(4) *For all nonnegative $\mu \in K_d$, the function*

$$x \mapsto E^x \int_0^{\tau_D} e^{A_s} dB_s$$

is bounded on D, where B_t is the positive continuous additive functional associated with μ.

(5) $\lambda_1 < 0$.

Proof. (1)$\Longrightarrow$(5)$\Longrightarrow$ (2)$\Longrightarrow$(1). For a given bounded function f on ∂D. we define for every $x \in D$,

$$\phi(x) = E^x\left\{e^{A_{\tau_D}} f(X(\tau_D)) 1_{\{\tau_D \le 1\}}\right\}.$$

From Theorem (2.5) we know that ϕ is bounded on D which implies $\phi \in L^2(D)$. By the Markov property we get

$$E^x\left\{e^{A_{\tau_D}} f(X(\tau_D))\right\} = \phi(x) + \sum_{n=1}^{\infty} T_n^D \phi(x). \tag{3.16}$$

for every $x \in D$. Suppose that f is strictly positive on ∂D and $E^x\{e^{A_{\tau_D}} f(X(\tau_D))\} < \infty$ for some $x \in D$, then (3.16) implies

$$\lim_{n \uparrow \infty} T_t^D \phi(x) = 0.$$

But it follows from the strict positivity of f that ϕ is strictly positive on D. Thus (1)$\Longrightarrow$(5) is proved by using Lemma (3.14).

Now suppose that $\lambda_1 < 0$. Let $\alpha = \frac{\lambda_1}{2}$. By the proof of Theorem (3.10) we get

$$\limsup_{t\uparrow\infty} \sup_{x\in D} e^{-\alpha t}|T_t^D\phi(x)| < \infty.$$

Therefore by (3.16),

$$\sup_{x\in D} E^x\left\{e^{A_{\tau_D}} f(X(\tau_D))\right\} \leq \|\phi\|_\infty + \sum_{n=1}^{\infty} e^{\alpha n} \sup_{x\in D} e^{-\alpha n}|T_n^D\phi(x)| < \infty,$$

which proves that (5)$\Longrightarrow$(2). (2)$\Longrightarrow$(1) is trivial.

(3)$\Longrightarrow$(5)$\Longrightarrow$(4) $\Longrightarrow$(3). For a nonnegative $\mu \in K_d$, we define

$$\phi(x) = E^x \int_0^{\tau_D\wedge 1} e^{A_s} dB_s$$

for every $x \in D$, then ϕ is bounded on D. By the Markov property we ahve

$$E^x \int_0^{\tau_D} e^{A_s} dB_s = \phi(x) + \sum_{n=1}^{\infty} T_n^D\phi(x),$$

for every $x \in D$. If μ is nonvanishing, then ϕ is strictly positive on D. Thus to complete the proof we use the same argument as in the first part of the proof.

In the remainder of this paper we are going to prove some important consequences of the above gauge theorem. These results will be used, in a subsequent paper, to solve the Dirichlet problem of the following equation

$$L\phi = 0,$$

where, as before, L stands for the generator of the semigroup $(T_t^D)_{t>0}$.

The first consequence that we are going to present is a result which may be called a supergauge theorem according to the terminolgy in [7]. The following lemma will be used to prove the super gauge theorem.

(3.17) Lemma. *If for any $r > 0$, we define*

$$\begin{aligned} U_t^{(D,r)} f(x) &= E^x\{e^{(1+r)A_t} f(X_t); t < \tau_D\} \\ &= E^x\{e^{A_t^{[(1+r)\mu]}} f(X_t); t < \tau_D\}. \end{aligned}$$

Then for any $t > 0$, $U_t^{(D,r)}$ converges to T_t^D strongly in $L^2(D)$ as $r \longrightarrow 0$.

Proof. Fix an arbitrary $t > 0$. For any $f \in L^2(D)$,

$$\begin{aligned}
&\sup_{x\in D} |U_t^{(D<r)} f(x) - T_t^D f(x)| \\
&= \sup_{x\in D} E^x\{(e^{(1+r)A_t} - e^{A_t}) f(X_t); t < \tau_D\}| \\
&\leq C_1 \|f\|_{L^2(D)} \sup_{x\in D} \left(E^x\{e^{2A_t}(e^{rA_t} - 1)^2\}\right)^{\frac{1}{2}} \\
&\leq C_1 \|f\|_{L^2(D)} \sup_{x\in D} \left(E^x e^{4A_t}\right)^{\frac{1}{4}} \sup_{x\in D} \left(E^x\{(e^{r|A_t|} - 1)^4\}\right)^{\frac{1}{4}} \\
&\leq C_2 \|f\|_{L^2(D)} \left(\sup_{x\in D} E^x e^{4r|A_t|} - 1\right)^{\frac{1}{4}},
\end{aligned}$$

where in the last line we used Theorem (2.12).

Since

$$\lim_{\eta\downarrow 0} \frac{e^\eta}{1 - 4\eta} = 1,$$

we know that for any $\epsilon > 0$ we can take an η with $0 < \eta < \frac{1}{4}$ such that

$$\frac{e^\eta}{1 - 4\eta} - 1 < \epsilon^4.$$

From the boundedness of u we know that there exists an $r_1 > 0$ such that $16r\|u\|_\infty < \eta$ whenever $0 < r < r_1$. Since $|\nabla u|^2$ is in the Kato class, we know that there exists an $r_2 > 0$ such that

$$\sup_{x\in R^d} \int_0^t 64r^2 |\nabla u|^2(X_s) ds < \eta^2$$

whenever $0 < r < r_2$. Repeating part of the argument in the proof of Lemma (2.1) we get that

$$\sup_{x\in R^d} E^x\{e^{8r|\int_0^t \nabla u(X_s)\cdot dX_s|}\} < \frac{1}{1 - 4\eta}$$

whenever $0 < r < r_2$. Take $r_3 = r_1 \wedge r_2$. Then

$$\begin{aligned}
&\sup_{x\in R^d} E^x\{e^{8r|\tilde{A}_t|}\} \\
&\leq e^\eta \sup_{x\in R^d} E^x\{e^{8r|\int_0^t \nabla u(X_s)\cdot dX_s|}\} \\
&\leq \frac{e^\eta}{1 - 4\eta} < 1 + \epsilon^4,
\end{aligned}$$

whenever $0 < r < r_3$. It follows from Khas'minskii's Lemma (see, for instance, [3]) that there exists an $r_4 > 0$ such that

$$\sup_x E^x\{e^{8r|\overline{A}_t|}\} \leq 1 + \epsilon^4$$

whenever $0 < r < r_4$. Take $r_0 = r_3 \wedge r_4$. Then we get

$$\sup_x E^x\{e^{4r|A_t|}\} \le \sup_x \left(E^x\{e^{8r|\tilde{A}_t|}\}\right)^{\frac{1}{2}} \sup_x \left(E^x\{e^{8r|\overline{A}_t|}\}\right)^{\frac{1}{2}}$$
$$\le 1 + \epsilon^4.$$

whenever $0 < r < r_0$. Therefore we have

$$\sup_{x\in D} |U_t^{(D,r)} f(x) - T_t^D f(x)| \le C_2 \epsilon \|f\|_{L^2(D)}$$

whenever $0 < r < r_0$. Consequently

$$\|U_t^{(D,r)} - T_t^D\|_{L^2(D),L^2(D)} \le C_3 \epsilon$$

whenever $0 < r < r_0$. The proof is now complete.

(3.18) Theorem(Super Gauge Theorem). *If the function*

$$g(x) = E^x\{e^{A_{\tau_D}}\}$$

is finite for some $x \in D$, then there is an r_0 such that

$$g_r(x) = E^x\{e^{(1+r)A_{\tau_D}}\}$$

is bounded on D whenever $0 < r < r_0$.

Proof. For each $r > 0$, let $\gamma_1^{(r)}$ be such that $e^{\gamma_1^{(r)}}$ is the largest eigenvalu of $U_1^{(D,r)}$. From Lemma (3.17) we know that $e^{\gamma_1^{(r)}} \longrightarrow e^{\lambda_1}$, or equivalently, $\gamma_1^{(r)} \longrightarrow \lambda_1$ as $r \longrightarrow 0$. Since $g(x)$ is finite for some $x \in D$, we know that $\lambda_1 < 0$ by Theorem (3.15). Therefore there exists an $r_0 > 0$ such that $\gamma_1^{(r)} < 0$ whenever $0 < r < r_0$. Hence by using Theorem (3.15) again we know that g_r is bounded on D whenever $0 < r < r_0$.

(3.19) Theorem. *If the function*

$$g(x) = E^x\{e^{A(\tau_D)}\}$$

is finite for some $x \in D$, then there is an N such that

$$g^{(n)}(x) = E^x\{e^{A^{(n)}(\tau_D)}\}$$

is bounded on D whenever $n \ge N$.

Before we prove Theorem (3.19), we first give an useful lemma.

(3.20) Lemma. *For each $t > 0$, $T_t^{(D,n)}$ converges to T_t^D strongly in $L^2(D)$.*

Proof. This proof is similar to that of Lemma (2.6). We give the complete proof here for the reader's convenience. Fix an arbitrary $t > 0$. For any $f \in L^2(D)$ and any relatively compact, open subset G having the closure of D as a subset,

$$\begin{aligned}
&\sup_{x\in D} E^x\{(e^{A_t} - e^{A_t^{(n)}})f(X_t); t < \tau_D\}\\
&\leq C_1\|f\|_{L^2(D)} \sup_{x\in D}\left(E^x\{e^{2A_t^{(n)}}|1 - e^{\tilde{A}_t - \tilde{A}_t^{(n)}}|^2 1_G(X_t)\}\right)^{\frac{1}{2}}\\
&\leq C_1\|f\|_{L^2(D)} \sup_{x\in D}\left(E^x e^{4A_t^{(n)}}\right)^{\frac{1}{4}} \sup_{x\in D}\left(E^x\{(1 - e^{\tilde{A}_t - \tilde{A}_t^{(n)}})^4 1_G(X_t)\}\right)^{\frac{1}{4}}\\
&\leq C_2\|f\|_{L^2(D)} \sup_{x\in D}\left(E^x\left\{(e^{|\tilde{A}_t - \tilde{A}_t^{(n)}|} - 1)^4 1_G(X_t)\right\}\right)^{\frac{1}{4}}\\
&\leq C_2\|f\|_{L^2(D)} \sup_{x\in D}\left(E^x\left\{e^{4|\tilde{A}_t - \tilde{A}_t^{(n)}|}1_G(X_t)\right\} - E^x 1_G(X_t)\right)^{\frac{1}{4}}\\
&\leq C_2\|f\|_{L^2(D)}\left(\sup_{x\in D} E^x\left\{e^{4|\tilde{A}_t - \tilde{A}_t^{(n)}|}1_G(X_t)\right\} - \inf_{x\in D} P^x(X_t \in G)\right)^{\frac{1}{4}}\\
&\leq C_2\|f\|_{L^2(D)}\left(\sup_{x\in D} E^x\left\{e^{4|\tilde{A}_t - A_t^{(n)}|}1_G(X_t) - 1\right\} +\right.\\
&\left. + \left\{1 - \inf_{x\in D} P^x(X_t \in G)\right\}\right)^{\frac{1}{4}},
\end{aligned}$$

where in the fourth line we used Lemma (2.4).

For any $\epsilon > 0$, take G so that

$$1 - \inf_{x\in D} P^x(X_t \in G) \leq \frac{\epsilon^4}{2C_2^4},$$

and then fix this G. Since

$$\lim_{\eta\downarrow 0}\frac{e^{8\eta}}{1 - 4\eta} = 1,$$

we can take an η with $0 < \eta < \frac{1}{4}$ such that

$$\frac{e^{8\eta}}{1 - 4\eta} < 1 + \frac{\epsilon^4}{2C_2^4}.$$

Now since $u_n \longrightarrow u$ uniformly on G, we know that there exists an N_1 such that

$$\sup_{x\in G}|u(x) - u_n(x)| < \eta,$$

whenever $n \geq N_1$. By Lemma (1.8) we know that there exists an N_2 such that

$$\sup_x E^x \int_0^T 16|\nabla(u-u_n)|^2(X_s)ds < \eta^2,$$

whenever $n \geq N_2$. Repeating part of the argument in the proof of Lemma (2.1), we get

$$\sup_x E^x \left\{e^{4|\int_0^t \nabla(u-u_n)(X_s)\cdot dX_s|}\right\} < \frac{1}{1-4\eta},$$

whenever $n \geq N_2$. Take $N = N_1 \vee N_2$. Then

$$\begin{aligned}&\sup_x E^x \left\{e^{4|\tilde{A}_t - \tilde{A}_t^{(n)}|} 1_G(X_t)\right\}\\ &\leq e^{8\eta} \sup_x E^x \left\{e^{4|\int_0^t \nabla(u-u_n)(X_s)\cdot dX_s|}\right\}\\ &\leq \frac{e^{8\eta}}{1-4\eta} < 1 + \frac{\epsilon^4}{2C_2^4},\end{aligned}$$

Therefore we have

$$\sup_{x\in D} E^x \left\{(e^{A_t} - e^{A_t^{(n)}})f(X_t); t < \tau_D\right\} \leq \epsilon \|f\|_{L^2(D)},$$

whenever $n \geq N$. Consequently

$$\|T_t^D - T^{(D,n)}\|_{L^2(D),L^2(D)} \leq C_3\epsilon$$

whenever $n \geq N$. The proof is now complete.

Now here is the proof of Theorem (3.19).

Proof of Theorem (3.19). For each $n \geq 1$, let $\lambda_1^{(n)}$ be such that $e^{\lambda_1^{(n)}}$ is the largest eigenvalue of $T_1^{(D,n)}$. From Lemma (3.18) we know that $e^{\lambda_1^{(n)}} \longrightarrow e^{\lambda_1}$, or equivalently, $\lambda_1^{(n)} \longrightarrow \lambda_1$. Since $g(x)$ is finite for some $x \in D$, we know that $\lambda_1 < 0$ by Theorem (3.15). Therefore there exists an N such that $\lambda_1^{(n)} < 0$ whenever $n \geq N$. Hence by using Theorem (3.15) again we know that $g^{(n)}$ is bounded on D whenever $n \geq N$.

From Theorem (3.19) we know that for any bounded measurable function f on ∂D the function

$$v^{(n)}(x) = E^x\{e^{A^{(n)}(\tau_D)} f(X_{\tau_D})\}$$

is bounded on D when n is large enough.

(3.21) Theorem. *The sequence of functions $v^{(n)}$ converges uniformly on D to the following function*

$$v(x) = E^x\{e^{A(\tau_D)}f(X_{\tau_D})\}.$$

Proof. Take an $M > 0$ so that $\frac{1}{M}$ is less than the r_0 specified in Theorem (3.18). Then

$$\begin{aligned}
&\sup_{x\in D} |v(x) - v^{(n)}(x)| \\
&\le \|f\|_\infty \sup_{x\in D} E^x\{e^{A(\tau_D)}(e^{|\tilde{A}(\tau_D)-\tilde{A}^{(n)}(\tau_D)|} - 1)\} \\
&\le \sup_{x\in D}\left(E^x\{e^{(1+\frac{1}{M})A(\tau_D)}\}\right)^{\frac{M}{M+1}} \sup_{x\in D}\left(E^x\{e^{(M+1)|\tilde{A}(\tau_D)-\tilde{A}^{(n)}(\tau_D)|} - 1\}\right)^{\frac{1}{M+1}} \\
&\le C\sup_{x\in D}\left(E^x\{e^{(M+1)|\tilde{A}(\tau_D)-\tilde{A}^{(n)}(\tau_D)|} - 1\}\right)^{\frac{1}{M+1}} \\
&\le C\left(\sup_{x\in D} E^x\{e^{(M+1)|\tilde{A}(\tau_D)-\tilde{A}^{(n)}(\tau_D)|}\} - 1\right)^{\frac{1}{M+1}}.
\end{aligned}$$

Since

$$\lim_{\eta\downarrow 0}\frac{e^\eta}{1-4\eta} = 1,$$

we know that for any $\epsilon > 0$ we can take an η with $0 < \eta < \frac{1}{4}$ such that

$$\frac{e^\eta}{1-4\eta} - 1 < \epsilon^{M+1}.$$

Now since $u_n \longrightarrow u$ uniformly on $\overline{D}$, we know that there exists an N_1 such that

$$2(M+1)\sup_{x\in\overline{D}}|u(x) - u_n(x)| < \eta,$$

whenever $n \ge N_1$. By Lemma (1.8) we know that there exists an N_2 such that

$$\sup_x E^x\int^{\tau_D}(M+1)^2|\nabla(u-u_n)|^2(X_s)ds < \eta^2,$$

whenever $n \ge N_2$. Repeating part of the argument in the proof of Lemma (2.1), we get

$$\sup_{x\in D} E^x\left\{e^{(M+1)|\int_0^{\tau_D}\nabla(u-u_n)(X_s)\cdot dX_s|}\right\} < \frac{1}{1-4\eta},$$

whenever $n \ge N_2$. Take $N = N_1 \vee N_2$. Then

$$\begin{aligned}
&\sup_{x\in D} E^x\{e^{(M+1)|\tilde{A}(\tau_D)-\tilde{A}^{(n)}(\tau_D)|}\} \\
&\le e^\eta \sup_{x\in D} E^x\left\{e^{(M+1)|\int_0^{\tau_D}\nabla(u-u_n)(X_s)\cdot dX_s|}\right\} \\
&\le \frac{e^\eta}{1-4\eta} < 1 + \epsilon^{M+1}
\end{aligned}$$

whenever $n \geq N$. Therefore we have

$$\sup_{x \in D} |v(x) - v^{(n)}(x)| \leq C\epsilon$$

whenever $n \geq N$. the proof is now complete.

REFERENCES

1. R. A. Adams, *Sobolev spaces*, Academic Press, New York, 1975.
2. N. Aizenman and B. Simon, *Brownian motion and Harnack inequality for Schrödinger operators*, Commun. Pure Appl. Math. **35**, 209–273.
3. Ph. Blanchard and Z. Ma, *Semigroup of Schrödinger operators with potentials given by radon measures*, Proc. 2nd Ascona, Locarno, Como Meeting (S. Albevrio, ed.), World Scientific Publications, Singapore (to appear).
4. ———, *New results on the Schrödinger semigroups with potentials given by signed smooth measures*, Lecture Notes in Math. **1444** (1990).
5. R. M. Blumenthal and R. K. Getoor, *Markov processes and potential theory*, Academic Press, New York, 1968.
6. K. L. Chung and K. M. Rao, *Feynman–Kac functional and the Schrödinger equation*, Seminar on Stochastic Processes (E. Cinlar, K. L. Chung and R. K. Getoor, eds.), Birkhauser, Boston, 1981.
7. ———, *General gauge theorem for multiplicative functionals*, Trans. Amer. Math. Soc. **306** (1988), 819–836.
8. C. Dellacherie and P. A. Meyer, *Probabilites et potentiel: théorie des martingales*, Herman, Paris, 1980.
9. M. Fukushima, *Dirichlet forms and Markov processes*, North–Holland, Amsterdam, 1980.
10. J. Glover, M. Rao and R. Song, *The gauge theorem fo a class of additive functionals of zero energy* (to appear).
11. Z. Ma, *Probabilistic treatment of Schrödinger equation with infinite gauge*, Scentia Sinica, **30** (1987), 685–695.
12. Z. Ma and R. Song, *Probabilistic methods in Schrödinger equations*, Seminar on Stochastic Processes (E. Cinlar, K. L. Chung and R. K. Getoor, eds.), Birkhauser, Boston, 1990.
13. B. Simon, *Schrödinger semigroups*, Bull. Amer. Math. Soc. **7** (1982), 447–536.

DEPARTMENT OF MATHEMATICS, UNIVERSITY OF FLORIDA, GAINESVILLE, FL 32611

On a limit theorem for non-stationary branching processes.

TETSUYA HATTORI and HIROSHI WATANABE

0. Introduction.

The purpose of this paper is to give a limit theorem for a certain class of discrete-time multi-type non-stationary branching processes (multi-type varying environment Galton-Watson processes). Let $X_N = {}^t(X_{N,1}, X_{N,2}, \cdots, X_{N,d})$, $N = 0, 1, 2, \cdots$ be a discrete-time branching process with d types. The discrete-time branching process X_N is determined by its generating functions $\phi_{n,N,i}(z)$. For $i \in \{1, 2, \cdots, d\}$, define $e_i = {}^t(e_{i,1}, e_{i,2}, \cdots, e_{i,d}) \in \mathbf{Z}_+{}^d$ by $e_{i,i} = 1$, and $e_{i,j} = 0$, if $i \neq j$. Then

$$\phi_{n,N,i}(z) \stackrel{d}{=} \sum_{\alpha \in \mathbf{Z}_+{}^d} \left(\prod_{i=1}^{d} z_i^{\alpha_i} \right) \mathbf{Prob}\{X_N = \alpha \mid X_n = e_i\}, \tag{1}$$

$$n \in \mathbf{Z}_+,\ N \in \mathbf{Z}_+,\ n \leq N,\ i = 1, 2, \cdots, d,\ z \in \mathbf{C}^d.$$

The generating functions $\phi_{n,N} = {}^t(\phi_{n,N,1}, \phi_{n,N,2}, \cdots, \phi_{n,N,d})$ are determined by recursion relations (see, for example, [2] on branching processes). For a non-stationary branching process, we have

$$\phi_{n,N}(z) = \phi_{n,N-1}(\phi_{N-1,N}(z)) = \phi_{n,n+1}(\phi_{n+1,N}(z)),\ n < N, \tag{2}$$

$$\phi_{n,n}(z) = z,$$

Compared with the stationary case, $\phi_{n,N} = \phi_{0,N-n}$, or the one-type non-stationary case, $d = 1$, not many studies seem to have been done for multi-type non-stationary branching processes. Perhaps one of the difficulties with such studies lies in finding a class of processes where the non-stationarity and multi-typedness enter in a non-trivial way, and yet a clear limiting behavior is obtained.

As an approach to this problem, we consider in this paper the case where the non-stationarity enters in a simple way. Namely, we assume

$$\phi_{N-1,N}(z) = D_{N-1}^{-1} F(D_N z),\ N = 1, 2, 3, \cdots, \tag{3}$$

where $D_N = \mathrm{diag}(D_{N,1}, D_{N,2}, \cdots, D_{N,d})$, $N \in \mathbf{Z}_+$, are diagonal matrices, and F is a $\mathbf{C}^d$-valued function in d variables.

From eq. (2) and eq. (3) it follows that

$$\phi_{n,N}(z) = D_n^{-1} F^{N-n}(D_N z),$$

which is similar to the corresponding formula for the stationary case. In fact, if the matrices D_N are unit matrices, this equation reduces to the formula for the stationary case.

If the limit $\hat{F}(z) = \lim_{N\to\infty} D_{N-1}^{-1} F(D_N z)$ exists and is analytic at $z = 1$, then the asymptotic behavior of the branching process will be essentially that of a stationary branching process defined by $\phi_{N-1,N} = \hat{F}$. However, if $\hat{F}$ is singular at $z = 1$, we cannot apply the theory of stationary branching processes directly, hence a consideration of non-stationarity (N-dependences of $\phi_{N-1,N}(z)$) becomes non-trivial.

We introduce some notation. Put $\vec{1} \stackrel{d}{=} {}^t(1, \cdots, 1)$. For a $\mathbf{C}^d$-valued differentiable function $F = {}^t(F_1, F_2, \cdots, F_d)$ in d variables $z = (z_1, \cdots, z_d)$, we define a $d \times d$ matrix $\nabla F(z)$ by $\nabla F(z)_{ij} = \dfrac{\partial F_i(z)}{\partial z_j}$. For a $d \times d$ matrix M, $\| M \| \stackrel{d}{=} \sup_{v=(v_1,\cdots,v_d),\ \|v\|=1} \| M v \|$ is the usual operator norm, with $\| v \| = \sqrt{\sum_{i=1}^{d} |v_i|^2}$ for a d-component vector v. We also use the notation $\exp(t) \stackrel{d}{=} (\exp(t_1), \cdots, \exp(t_d))$ for $t = (t_1, \cdots, t_d) \in \mathbf{C}^d$.

Definition 1. Let $\{D_N\}$, $N = 0, 1, 2, \cdots$, be a series of d-dimensional diagonal matrices, $D_N = \mathrm{diag}(D_{N,1}, D_{N,2}, \cdots, D_{N,d})$, with positive diagonal elements; $D_{N,i} > 0$. For a constant $\delta > 1$, we say that $\{D_N\}$ is of order δ if $\sup_{N\in\mathbf{Z}_+, i=1,\cdots,d} D_{N,i} < \infty$ and $\sup_{N\in\mathbf{Z}_+, i=1,\cdots,d} \delta^{-N} D_{N,i}^{-1} < \infty$ are satisfied.

Definition 2. Let $\{D_N\}$, $N = 0, 1, 2, \cdots$, be a series of d-dimensional diagonal matrices of order δ (> 1) and let ℓ be a constant satisfying $\ell > \delta$. We say that a $\mathbf{C}^d$-valued function F in d variables defined on an open set containing $\bigcup_{N\in\mathbf{Z}_+} \{D_N \vec{1}\}$ is $(\{D_N\}, \ell)$-regular if the following are satisfied.

1. The function F has the expression

$$F(D_N \vec{1} + t) = D_{N-1}\vec{1} + A_N t + \tilde{F}_N(t), \quad N \in \mathbf{Z}_+, \ \| t \| < C_0\, \delta^{-N},$$

   where A_N is a $d \times d$ matrix independent of t, and $\tilde{F}_N$ is analytic in $\| t \| < C_0\, \delta^{-N}$ and satisfies

$$\| \tilde{F}_N(t) \| \le C_1\, \delta^N \| t \|^2, \quad \| t \| < C_0\, \delta^{-N}, \ N \in \mathbf{Z}_+,$$

   for some positive constants C_0 and C_1 independent of N and t.
2. There exists a matrix A such that

$$\| A - A_N \| < C_2\, \delta^{-N},$$

where C_2 is a constant independent of N. The matrix A has an eigenvalue ℓ, and the absolute values of eigenvalues of A other than ℓ are less than ℓ. Furthermore, each Jordan cell with the eigenvalue ℓ is one dimensional.

The condition 1 implies that

$$F(D_N \vec{1}) = D_{N-1} \vec{1}\,,\ N = 1, 2, 3, \cdots,$$
$$A_N = \nabla F(D_N \vec{1})\,,\ N = 1, 2, 3, \cdots,$$

and the condition 2 implies that the limit

$$B_n \stackrel{d}{=} \lim_{N\to\infty} \ell^{-N+n} A_{n+1} A_{n+2} \cdots A_N \tag{4}$$

exists (see Appendix).

The conditions 1 and 2 roughly state that for each $N \in \mathbf{Z}_+$, the function F is analytic in a circle of center $D_N \vec{1}$ and radius $O(\delta^{-N})$ and that the first order term of F in z dominates the higher order terms.

Note that the conditions do not exclude the possibility that the function F is singular at the point $z_0 \stackrel{d}{=} \lim_{N\to\infty} D_N \vec{1}$.

Remark. We can extend our results to the case where the function F has N-dependences; i.e. with the replacement $F \to F_N$ in the above definition and in eq. (3), Theorem 1 and 2 below still hold.

Definition 3. Let $\{D_N\}$, $N = 0, 1, 2, \cdots$, be a series of d-dimensional diagonal matrices of order $\delta\,(> 1)$. We call X_N a branching process of type $(d, \{D_N\}, \ell)$, if X_N is a discrete-time branching process with d types, and if the generating functions defined by eq. (1) satisfy the recursion equations eq. (2) and eq. (3) for some $(\{D_N\}, \ell)$-regular function F.

The conditions in Definition 1 and Definition 2.2 together with $\ell > \delta$ are an extension of the notion of supercriticality. Definition 2.1, on the other hand, is a rather restrictive condition implying the existence of moments of X_N.

In this paper we prove the following.

Theorem 1. *Let X_N be a branching process of type $(d, \{D_N\}, \ell)$. Then $\ell^{-N} D_N^{-1} X_N$ converges weakly to a random variable $Y = (Y_1, \cdots, Y_d)$ as $N \to \infty$.*

$$E[Y_j \mid X_n = e_i] = \ell^{-n} \left(D_n^{-1} B_n\right)_{ij}\,,$$
$$i = 1, 2, \cdots, d\,,\ j = 1, 2, \cdots, d\,,\ n \in \mathbf{Z}_+\,.$$

Example. Let X_N be a branching process with 2 types such that the transition probability

$$P_N(\alpha,\beta) = \mathbf{Prob}\{X_{N+1} = \beta \mid X_N = \alpha\}, \quad \alpha \in \mathbf{Z}_+{}^2, \ \beta \in \mathbf{Z}_+{}^2, \ N \in \mathbf{Z}_+,$$

is determined by

$$\begin{aligned} P_N((1,0),(2,0)) &= \frac{(1-\epsilon'_{N+1})^2}{1-\epsilon'_N}, \\ P_N((1,0),(k,2)) &= \frac{3(1-\epsilon'_{N+1})^k \epsilon_{N+1}^2}{25(1-\epsilon'_N)}, \quad k = 0,1,2,\cdots, \\ P_N((1,0),\beta) &= 0, \text{ otherwise}, \\ P_N((0,1),(k,2)) &= \frac{(1-\epsilon'_{N+1})^k \epsilon_{N+1}^2}{\epsilon_N}, \quad k = 0,1,2,\cdots, \\ P_N((0,1),\beta) &= 0, \text{ otherwise}. \end{aligned}$$

Here, ϵ'_N, ϵ_N, $N = 0,1,\cdots$, are positive constants, defined recursively by $F(1-\epsilon'_{N+1}, \epsilon_{N+1}) = (1-\epsilon'_N, \epsilon_N)$, $N \in \mathbf{Z}_+$, with given positive constants ϵ'_0 and ϵ_0, satisfying $\epsilon'_0 < \frac{1}{2}$ and $\epsilon_0 < \frac{1}{2}$, and F is defined by

$$F(x,y) = (x^2 + \frac{3}{25}\frac{y^2}{1-x}, \frac{y^2}{1-x}).$$

It is easy to see that X_N is a branching process of type $(2,\{D_N\},4)$ with $D_N = \mathrm{diag}(1-\epsilon'_N, \epsilon_N)$. The conditions in the definitions are satisfied with $\delta = \frac{5}{3}$. Also it follows that $\epsilon_N = C\,\delta^{-N}(1+O(\delta^{-N}))$ and $\epsilon'_N = C\,\delta^{-N-1}(1+O(\delta^{-N}))$, where C is a constant. Theorem 1 therefore implies that $4^{-N}((1-\epsilon'_N)^{-1} X_{N,1}, \epsilon_N^{-1} X_{N,2})$ converges weakly as $N \to \infty$. Using the asymptotic form of ϵ'_N and ϵ_N, it further follows that $4^{-N} X_{N,1}$ and $(12/5)^{-N} X_{N,2}$ converges, or in other words, the average number of type 1 will grow at the rate of 4 per generation asymptotically, while that of type 2 will grow at the rate of 12/5 per generation.

Note that though the birth rate of type 2 from type 1 disappears in the limit $N \to \infty$, the birth affects the average growth rate significantly in the limit, or otherwise the average growth rate for type 1 would have been 2 instead of $\ell = 4$. This is because a birth of type 2 will in turn give birth of order $\frac{1}{\epsilon'_N}$ of type 1 in average, which diverges as N is increased.

For the one-type case $d = 1$, the branching process of type $(d = 1, \{D_N\}, \ell)$ falls into the class considered by Biggins and D'Souza [3]. In this case, the conditions in Definition 1 and Definition 2.2 together with $\ell > \delta$ imply the "uniform supercriticality", while Definition 2.1 implies the "dominance condition". (We learned this from Prof. Biggins.) A result in [3] then implies that (for $d = 1$) the process has a single rate of growth.

As is seen in the above example, the conditions (for $d > 1$) do *not* imply the same rate of growth *among the different types.* The gap between the upper and lower bound of $D_{N,i}$ in Definition 1 is crucial for such a possibility.

We are particularly interested in the situation where growth rate of different types differ, and where the birth of less increasing types affects the growth rates of dominant types. Non-stationarity and multi-typedness both enter in a non-trivial way in this phenomena.

The present study in the particular class of branching processes grew out of our attempt in constructing a non-self-similar diffusion on the Sierpinski gasket. We briefly discuss the application in Section 3 (the details, however, will be left for the future publication).

Acknowledgement. The authors would like to thank Prof. M. T. Barlow, Prof. K. Hattori, Prof. J. Kigami, Prof. T. Kumagai, Prof. S. Kusuoka, and Prof. Y. Takahashi for valuable discussions concerning the present work. They would also like to thank Prof. J. D. Biggins who kindly informed them of the connection between the present work and his results on one-type varying environment Galton-Watson processes.

1. Main results.

Let $\{\phi_{n,N}(z)\}$ be a sequence of $\mathbf{C}^d$-valued functions defined by the recursion relation eq. (2) and eq. (3) for a $\mathbf{C}^d$-valued $(\{D_N\},\ell)$-regular function F. Put

$$\begin{aligned} f_{n,N}(s) &= D_n\phi_{n,N}(\exp(-\ell^{-N}D_N^{-1}s)) \\ &= F^{N-n}(D_N\exp(-\ell^{-N}D_N^{-1}s)) \,. \end{aligned}$$

Theorem 2. *There exist positive constants ϵ and C such that, for every $n \in \mathbf{Z}_+$:*

(i) The function $f_{n,N}(s)$, $N \geq n$, is analytic on $\| s \| < \epsilon \ell^n \delta^{-n}$ and converges uniformly to an analytic function $f_n(s)$ on $\| s \| < \epsilon \ell^n \delta^{-n}$ as $N \to \infty$.

(ii) The function $f_n(s)$ has the expression

$$f_n(s) = D_n\vec{1} - \ell^{-n}B_n\, s + \tilde{f}_n(s)\,, \quad \| s \| < \epsilon \ell^n \delta^{-n}\,,$$

with the bound

$$\| \tilde{f}_n(s) \| \leq C\,\delta^n \ell^{-2n} \| s \|^2, \quad \| s \| < \epsilon \ell^n \delta^{-n}\,.$$

Theorem 1 is obtained from Theorem 2 by a standard argument.

Proof of Theorem 1 assuming Theorem 2. Put $\mathbf{C}_+ \stackrel{d}{=} \{s \in \mathbf{C} \mid \Re(s) \geq 0\}$, For each $n \in \mathbf{Z}_+$ and $N \in \mathbf{Z}_+$, satisfying $n \leq N$, and for each $i \in \{1, 2, \cdots, d\}$, let $Q_{n,N,i}$ be the distribution of $\ell^{-N} D_N^{-1} X_N$ conditioned by $X_n = e_i$, and let

$$\begin{aligned}\Phi_{n,N,i}(s) &= \int_{\mathbf{R}^d} \exp(-s \cdot w)\, Q_{n,N,i}(dw) \\ &= \int_{[0,\infty)^d} \exp(-s \cdot w)\, Q_{n,N,i}(dw), \quad s \in {\mathbf{C}_+}^d,\end{aligned}$$

be the Laplace transform (generating function) of $Q_{n,N,i}$. Put $\Phi_{n,N} = (\Phi_{n,N,1}, \cdots, \Phi_{n,N,d})$. With eq. (1) we have

$$\Phi_{n,N}(s) = \phi_{n,N}(\exp(-\ell^{-N} D_N^{-1} s)).$$

Theorem 2 then implies that there exists a positive constant ϵ such that $\{\Phi_{n,N}\}_{N=n,n+1,n+2,\cdots}$ converges uniformly to an analytic function on $\| s \| < \epsilon \ (\leq \epsilon\, \ell^n \delta^{-n})$. Therefore we have for each $i \in \{1, 2, \cdots, d\}$ and each $n \in \mathbf{Z}_+$,

$$M \stackrel{d}{=} \sup_N \{ \int_{[0,\infty)^d} \exp(\epsilon \vec{1} \cdot w / (2\sqrt{d})) Q_{n,N,i}(dw) \} < \infty.$$

Fix $R > 0$ and put $E_j \stackrel{d}{=} \{w \in [0, \infty)^d \mid w_j \geq R\}$, $j = 1, 2, \cdots, d$. Then we have

$$\begin{aligned}M &\geq \sup_N \{ \int_{E_j} \exp(\epsilon \vec{1} \cdot w / (2\sqrt{d})) Q_{n,N,i}(dw) \} \\ &\geq \exp(\epsilon R / (2\sqrt{d})) \sup_N Q_{n,N,i}(E_j).\end{aligned}$$

Hence

$$\begin{aligned}\sup_N Q_{n,N,i}([0, R]^d) &\geq 1 - \sum_j \sup_N Q_{n,N,i}(E_j) \\ &\geq 1 - M\, d \exp(-\epsilon R / (2\sqrt{d})).\end{aligned}$$

The right hand side converges to 1 as $R \to \infty$. Therefore the sequence of measures $\{Q_{n,N,i}\}_{N=n,n+1,\cdots}$ is tight.

Since $\{Q_{n,N,i}\}$ is tight, there exists a subsequence that converges weakly to a probability measure. For each convergent subsequence $\{Q_{n,k_N,i}\}_{N=1,2,\cdots}$, $\{\Phi_{n,k_N,i}(s)\}$ converges uniformly on compact sets in

$\mathbf{C}_+{}^d$. Therefore $\{\Phi_{n,k_N,i}(s)\}$ converges to a function that is analytic in the interior of $\mathbf{C}_+{}^d$ and continuous in $\mathbf{C}_+{}^d$.

On the other hand, Theorem 2 implies that $\{\Phi_{n,N,i}(s)\}$ converges to an analytic function in a neighborhood of $s = 0$. Therefore, $\{\Phi_{n,k_N,i}\}$ converges to an analytic function independent of the choice of subsequences in the interior of $\mathbf{C}_+{}^d$. Since the limit of $\{\Phi_{n,k_N,i}\}$ is continuous in $\mathbf{C}_+{}^d$, this further implies that $\{\Phi_{n,k_N,i}\}$ converges to a continuous function independent of the choice of subsequences in $\mathbf{C}_+{}^d$. We have shown that for any convergent subsequence $\{Q_{n,k_N,i}\}$ of $\{Q_{n,N,i}\}$ the corresponding sequence of the characteristic functions $\{\Phi_{n,k_N,i}(\sqrt{-1}x)\}$ converges to a function independent of the choice of subsequences for $x \in \mathbf{R}^d$, which implies that the limit measure $Q_{n,i}$ of the sequence $\{Q_{n,k_N,i}\}$ is independent of the subsequence. Since $\{Q_{n,N,i}\}$ is tight, this implies that $\{Q_{n,N,i}\}$ converges weakly to a probability measure $Q_{n,i}$. The proof of the statement on the expectation values is straightforward. This completes the proof. ∎

2. Proof of Theorem 2.

We put, for $n_1, n_2 \in \mathbf{Z}_+$ with $n_1 \leq n_2$,

$$B_{n_1,n_2} \stackrel{d}{=} \ell^{-n_2+n_1} A_{n_1+1} A_{n_1+2} \cdots A_{n_2}, \tag{5}$$

$$B_{n_1,n_1} \stackrel{d}{=} I. \tag{6}$$

Theorem 2 follows from Propositions 1 and 2 below.

Proposition 1. *There exist positive constants ϵ and C_3 independent of $n, N \in \mathbf{Z}_+$ such that the function $f_{n,N}(s)$, $n \leq N$, has the expression*

$$f_{n,N}(s) = D_n \vec{1} - \ell^{-n} B_{n,N} s + \tilde{f}_{n,N}(s), \quad \| s \| < 2\epsilon\, \ell^n\, \delta^{-n}, \tag{7}$$

with the bound

$$\| \tilde{f}_{n,N}(s) \| \leq C_3 \delta^n \ell^{-2n} \| s \|^2, \quad \| s \| < 2\epsilon\, \ell^n\, \delta^{-n}. \tag{8}$$

Proposition 2. *For $n, N, N' \in \mathbf{Z}_+$ with $n \leq N \leq N'$, it holds that*

$$\| \tilde{f}_{n,N}(s) - \tilde{f}_{n,N'}(s) \| \leq C_4 r^{-N} \| s \|^2, \quad \| s \| < \epsilon\, \ell^n\, \delta^{-n}, \tag{9}$$

where $C_4 > 0$ and $r > 1$ are constants independent of n, N, N' and s.

Proof of Theorem 2 assuming Propositions 1 and 2. Proposition 2 implies that the family $\{\tilde{f}_{n,N}\}_{N=n,n+1,\cdots}$ constitutes a Cauchy series. Then, taking the limit $N \to \infty$ in (7) and (8), we obtain Theorem 2, where

$$\tilde{f}_n(s) = \lim_{N\to\infty} \tilde{f}_{n,N}(s). \quad ∎$$

Proof of Proposition 1. Let us prove the proposition by induction on $n = N, N-1, \cdots, 0$. Since

$$f_{N,N}(s) = D_N \exp(-\ell^{-N} D_N^{-1} s) \tag{10}$$

and since there exists a constant $C_5 > 0$ such that

$$\| \ell^{-N} D_N^{-1} s \| < C_5, \quad \| s \| < 2\epsilon \ell^N \delta^{-N},$$

we have

$$\| \tilde{f}_{N,N}(s) \| \leq \alpha \delta^N \ell^{-2N} \| s \|^2, \quad \| s \| < 2\epsilon \ell^N \delta^{-N}, \tag{11}$$

for some constant α independent of N and s.

Suppose that

$$\| \tilde{f}_{n,N}(s) \| \leq \alpha_n \delta^n \ell^{-2n} \| s \|^2, \quad \| s \| < 2\epsilon \ell^n \delta^{-n}, \tag{12}$$

holds for an $n \leq N$, where α_n is a constant independent of N and s. Let us estimate the function $f_{n-1,N}(s)$ given by

$$f_{n-1,N}(s) = F(f_{n,N}(s)) = F(D_n \vec{1} - \ell^{-n} B_{n,N} s + \tilde{f}_{n,N}(s)). \tag{13}$$

Firstly, the function $f_{n-1,N}(s)$ is well-defined by (13) for s such that

$$\| \ell^{-n} B_{n,N} s - \tilde{f}_{n,N}(s) \| < C_0 \delta^{-n}. \tag{14}$$

As is shown in Appendix, there exists a constant $C_6 > 0$ independent of n such that

$$\| B_{n,N} \| < C_6. \tag{15}$$

Then the condition (14) holds for s satisfying

$$\| s \| < 2\epsilon \ell^{n-1} \delta^{-n+1}, \tag{16}$$

if

$$C_6 \frac{\delta}{\ell} 2\epsilon + 4\epsilon^2 \left(\frac{\delta}{\ell}\right)^2 \alpha_n < C_0. \tag{17}$$

Secondly, by using

$$\tilde{f}_{n-1,N}(s) = A_n \tilde{f}_{n,N}(s) + \tilde{F}_n(-\ell^{-n} B_{n,N} s + \tilde{f}_{n,N}(s))$$

and

$$\| \ell^{-n} B_{n,N} s - \tilde{f}_{n,N}(s) \| \leq (C_6 + 2\epsilon\alpha_n)\ell^{-n} \| s \|,$$

we can estimate $\tilde{f}_{n-1,N}(s)$ and reproduce (12) with n replaced by $n-1$, where we put

(18) $$\alpha_{n-1} = \| A_n \| \delta\ell^{-2}\alpha_n + C_1\delta\ell^{-2}(C_6 + 2\epsilon\alpha_n)^2.$$

Since $\delta < \ell$ and

$$\| A_n \| < \ell + C_2\delta^{-n},$$

the solution to (18) with $\alpha_N = \alpha$ satisfies (17), if ϵ is sufficiently small. ∎

The proof of Proposition 2 is based on the recursion relation

(19) $$f_{m-1,N}(s) - f_{m-1,N'}(s) = F(f_{m,N}(s)) - F(f_{m,N'}(s))$$

and on the following difference property for F.

Lemma . *For $m \in \mathbf{Z}_+$ and $t, t' \in \mathbf{C}^d$ such that $\| t \|, \| t' \| < \frac{C_0}{2}\delta^{-m}$,*

(20) $$F(D_m\vec{1} + t) - F(D_m\vec{1} + t') = (A_m + G_m)(t - t'),$$

where $G_m = G_m(t,t')$ obeys the bound

(21) $$\| G_m \| \leq C_7\delta^m \max(\| t \|, \| t' \|).$$

Proof. The matrix element of $A_m + G_m$ is given by the corresponding element of ∇F at some point on the line segment connecting $D_m\vec{1} + t$ and $D_m\vec{1} + t'$. Note that

$$\nabla F(D_m\vec{1} + t'') = A_m + \nabla\tilde{F}_m(D_m\vec{1} + t'')$$

holds and that $\tilde{F}_m(D_m\vec{1} + t'')$ is analytic on $\{t'' \in \mathbf{C}^d | \; \| t'' \| < C_0\delta^{-m}\}$ and bounded by $C_1\delta^m \| t'' \|^2$. Now, we fix t'' such that $\| t'' \| < \frac{C_0}{2}\delta^{-m}$ and define the function $v(\tau) = \tilde{F}_m(D_m\vec{1} + t'' + \tau e_i)$ of τ analytic on $\{\tau \in \mathbf{C} | \; |\tau| < \frac{C_0}{2}\delta^{-m}\}$, where

$$e_i = {}^t(e_{i,1}, e_{i,2}, \cdots, e_{i,d}),$$
$$e_{i,j} = \delta_{i,j}.$$

Then, integrating the function $v(\tau)\tau^{-2}$ along the contour $|\tau| = \| t'' \|$, we can bound $v'(0) = \frac{\partial}{\partial t''_i}\tilde{F}_m(D_m\vec{1} + t'')$. As a result, we have

$$\| \frac{\partial}{\partial t''_i}\tilde{F}_m(D_m\vec{1} + t'') \| \leq \sup_{|\tau|=\|t''\|} \frac{\| v(\tau) \|}{\| t'' \|}$$
$$\leq 4C_1\delta^m \| t'' \|, \quad \| t'' \| < \frac{C_0}{2}\delta^{-m}.$$

This proves the lemma. ∎

Proof of Proposition 2. Fixing n, N, N' and s such that $n \leq N \leq N'$ and $\| s \| < \epsilon\delta^{-n}\ell^n$, we estimate $\| \tilde{f}_{m,N}(s) - \tilde{f}_{m,N'}(s) \|, m = N, N-1, \cdots, n$.

As is shown in Appendix, there exist positive constants $C_8, r' > 0$ with $1 < r' < \ell$ such that

$$\| B_{m,N} - B_{m,N'} \| < C_8 r'^{-N+m}, \quad m \leq N < N'. \tag{22}$$

Now, (7), (19), and (20) imply

$$\begin{aligned} \tilde{f}_{m-1,N}(s) - \tilde{f}_{m-1,N'}(s) &= F(f_{m,N}(s)) - F(f_{m,N'}(s)) + \ell^{-m+1}(B_{m-1,N} - B_{m-1,N'})s \\ &= (A_m + G_m)(f_{m,N}(s) - f_{m,N'}(s)) + \ell^{-m+1}(B_{m-1,N} - B_{m-1,N'})s \\ &= (A_m + G_m)(\tilde{f}_{m,N}(s) - \tilde{f}_{m,N'}(s)) + G_m\ell^{-m}(B_{m,N} - B_{m,N'})s, \end{aligned} \tag{23}$$

where G_m depends on s. By the argument obtaining (14) from (15), (16), and (17) with the ϵ replaced by $\frac{\epsilon}{2}$, we see that t and t' defined by

$$t = -D_m\vec{1} + f_{m,N}(s) = \ell^{-m}B_{m,N}s - \tilde{f}_{m,N}(s)$$
$$t' = -D_m\vec{1} + f_{m,N'}(s) = \ell^{-m}B_{m,N'}s - \tilde{f}_{m,N}(s)$$

satisfy the assumption of Lemma. Then we have the bound

$$\| G_m \| \leq C_7(C_6 + C_3\epsilon)\delta^m\ell^{-m} \| s \|,$$

since (8) and (15) imply

$$\| t \|, \| t' \| \leq (C_6 + C_3\epsilon)\ell^{-m} \| s \| .$$

Then, from (23), (15), and (22), we obtain

$$\begin{aligned} \| \tilde{f}_{m-1,N}(s) - \tilde{f}_{m-1,N'}(s) \| &\leq \| A_m + G_m \| \| \tilde{f}_{m,N}(s) - \tilde{f}_{m,N'}(s) \| \\ &\quad + C_7C_8(C_6 + C_3\epsilon)\delta^m\ell^{-2m}r'^{-N+m} \| s \|^2 . \end{aligned} \tag{24}$$

On the other hand, (8) implies

$$\| \tilde{f}_{N,N}(s) - \tilde{f}_{N,N'}(s) \| \leq \| \tilde{f}_{N,N}(s) \| + \| \tilde{f}_{N,N'}(s) \| \leq 2C_3\delta^N \ell^{-2N} \| s \|^2 . \tag{25}$$

The inequalities (24) and (25) together with the bound

$$\| A_m + G_m \| < \ell + C_2\delta^{-m} + C_7(C_6 + C_3\epsilon)(\frac{\delta}{\ell})^m \| s \|$$
$$< \ell + C_2\delta^{-m} + C_7\epsilon(C_6 + C_3\epsilon)(\frac{\delta}{\ell})^{m-n}$$

prove (9) with $r = \min(r', \ell/\delta)$. ∎

3. An application.

In this section, we briefly explain an application of our theorems to the construction of "asymptotically one-dimensional" diffusion processes on fractals. (The idea is announced in [5]. Details are in preparation.) Unlike the other part of this paper, we assume in this section that the reader is familiar with the works on diffusion processes on finitely ramified fractals (for example, [1][4][7][8][9]).

In the construction of diffusion processes on the finitely ramified fractals, one starts with a set of random walks on pre-fractals and obtains the diffusion process on the fractal as a weak limit of the (time-scaled) random walks. This approach was first established for the Sierpinski gasket [4][8][1]. One of the keys to this construction is decimation invariance, which means that the transition probability of the random walk is a fixed point of the corresponding renormalization group transformations (in the terminology of [6;section4.9]) in the space of transition probabilities. In general, one may assign different transition probabilities to different kinds of jumps of the random walk, in which case the space of transition probabilities become multi-dimensional. The condition that the obtained diffusion process spans the whole fractal implies that every component of the fixed point is positive (non-degenerate fixed point). In fact, Lindstrom defined a class of fractals (nested fractals) in such a way that the corresponding renormalization group has a non-degenerate fixed point [9;section IV, V], and succeeded in constructing diffusion processes on nested fractals.

In [6;section 5.4] we introduced *abc*-gaskets as examples of the fractals where non-degenerate fixed points of the corresponding renormalization groups are absent (if the parameters a, b, c satisfy certain conditions), while they always have unstable degenerate fixed points which correspond to the random walks on (one-dimensional) chains. The problem then arises; can we construct diffusion processes on finitely ramified fractals whose renormalization groups have only degenerate fixed points?

The idea of the solution is to use the renormalization group trajectories that "emerge" from a neighborhood of the unstable degenerate fixed points. We take the 111-gasket (which is just the Sierpinski gasket), as an example to explain briefly our idea of the construction, with emphasis on how the problem is related to non-stationary branching processes.

Let G_n denote the set of vertices of the pre-Sierpinski gasket with the smallest unit being the equilateral triangle of side length 2^{-n} (see, for example, [1] for the notation), and let $G = \overline{\bigcup_{n=0}^{\infty} G_n}$ denote the Sierpinski gasket. For a process X taking values in G we define a stopping time $T_i^n(X)$, $n = 0, 1, 2, \cdots$, by $T_0^n(X) = \inf\{t \geq 0 \mid X(t) \in G_n\}$, and $T_{i+1}^n(X) = \inf\{t > T_i^n(X) \mid X(t) \in G_n \setminus \{X(T_i^n(X))\}\}$ for $i = 0, 1, 2, \cdots$. $T_i^n(X)$ is the time that X hits G_n for the i-th time, counting only once if it hits the same point more than once in a row. Denote by $W^n(X) = T_1^n(X) - T_0^n(X)$ the time interval to hit two points in G_n. For an integer n and a process X on G or on G_N for some $N > n$, decimation is an operation that assigns a walk X' on G_n defined by $X'(i) = X(T_i^n(X))$.

The constructions in [4][8][1] use the sequence of simple random walks; random walks with same transition probability in every direction. The sequence of simple random walks $\{Y_N\}$ on G_N ($N = 1, 2, 3 \cdots$) has the property of decimation invariance; namely, the random walk Y' on G_n defined by $Y'(i) = Y_N(T_i^n(Y_N))$ has the same law as Y_n. This implies that $\{W^n(Y_N)\}$, $N = n+1, n+2, n+3, \cdots$, is a (one-type) *stationary* branching process (see, for example, [1;Lemma2.5(b)]). A limit theorem for a stationary branching process gives a necessary estimate, which, together with other ingredients, finally leads to the theorem that the sequence of processes X_N, $N = 1, 2, 3, \cdots$, defined by $X_N(t) = Y_N([\lambda^{-N} t])$, where $\lambda = E[W^0(Y_1)]$, converges weakly to a diffusion process as $N \to \infty$. The situation is similar for the case of multi-dimensional parameter space, which appears in nested fractals [8], where the limit theorems for multi-type stationary branching processes can be employed for necessary estimates [7].

As a generalization of the simple random walk let us consider a random walk $Z = Z_{N,\vec{x}}$, $\vec{x} = {}^t(x, y, z)$, on G_N defined as follows. Z is a Markov chain taking values on G_N, and the transition probability is defined in such a way that at each integer time Z jumps to one of the four neighboring points and the relative rate of the jump is $x : y : z$ for {a horizontal jump} : {a jump in 60° (or − 120°) direction} : {a jump in 120° (or −60°) direction}. For a simple random walk, $x = y = z$. The parameter space P can be defined as $P = \{(x, y, z) \mid x + y + z = 1,\ x \geq 0,\ y \geq 0,\ z \geq 0\}$. There is a one to one correspondence between a point in P and a random walk.

$Z_{N,\vec{x}}$ does not have the property of decimation invariance. Instead, the decimated walk Z' defined by $Z'(i) = Z_{N,\vec{x}}(T_i^{N-1}(Z_{N,\vec{x}}))$ has the same law as $Z_{N-1,\vec{x}'}$, where $\vec{x}' = T\vec{x} = {}^t(C(x + yz/3),\ C(y + zx/3),\ C(z +$

$xy/3))$, $1/C = 1 + (xy + yz + zx)/3$. The map T maps P into P. The dynamical system on the parameter space P defined by T is the renormalization group. The simple random walk corresponds to $(1/3, 1/3, 1/3) \in P$ which is a non-degenerate fixed point of T. $(0,0,1) \in P$ is a degenerate fixed point corresponding to the random walk along a one-dimensional chain.

We choose a sequence $\{Z_{N,\vec{x}_N}\}$ as follows. Let $0 < w_0 < 1$ and define w_N, $N = 1,2,3,\cdots$, inductively by $w_{N+1} = (6 - w_N)^{-1} \left(-2 + 3w_N + (4 + 6w_N + 6w_N^2)^{1/2}\right)$. Note that $w_0 > w_1 > w_2 > \cdots \to 0$. Put $\vec{x}_N = (1 + 2w_N)^{-1} (w_N, w_N, 1)$ and consider $Z_N = Z_{N,\vec{x}_N}$. One then sees that if $n < N$, then Z_n is a decimation of Z_N. This property of $\{Z_N\}$ corresponds to the decimation invariance of simple random walks $\{Y_N\}$. The special choice of the parameters simply means $\vec{x}_{N-1} = T\vec{x}_N$. Letting $N \to \infty$, $\vec{x}_N$ approaches the degenerate unstable fixed point $(0,0,1)$. Put $W_1^n(Z_N) = W^n(Z_N)$, and let $W_2^n(Z_N)$ be the number of diagonal (off-horizontal) jumps in the time interval $(T_0^n(Z_N),\ T_1^n(Z_N)]$, and $W_3^n(Z_N)$ be the number of visit in the same time interval to the points from which two horizontal lines emerges. Let D_N, $N = 0,1,2,\cdots$, be a sequence of diagonal matrices defined by $D_N = diag\left(\frac{1}{1+3w_N},\ w_N,\ \frac{1+3w_N}{2+2w_N}\right)$ Then by explicit calculations one obtains the following Proposition.

Proposition 3. $W_N = {}^t(W_1^n(Z_{n+N}),\ W_2^n(Z_{n+N}),\ W_3^n(Z_{n+N}))$, $N = 0,1,2,\cdots$, *is a branching process of type* $(d = 3, \{D_N\}, \ell = 6)$.

With this proposition, Theorems 1 and 2 can be applied, which take a part of the role of the limit theorems for stationary branching processes for the fixed point theories. Together with other ingredients we finally obtain the theorem that the sequence of processes X_N, $N = 1,2,3,\cdots$, defined by $X_N(t) = Z_N([\lambda^{-N}t])$, where $\lambda = \ell$, converges weakly to a diffusion process as $N \to \infty$. (The resulting process is different from the already known diffusion process on the Sierpinski gasket.) A Proposition similar to the Proposition 3 holds for general *abc*-gaskets, hence the Theorems 1 and 2 are generally applicable.

In contrast to the decimation invariant (fixed point) theories, where the existing limit theorems for the stationary branching processes worked, we needed limit theorems for the non-stationary multi-type branching processes. The present study grew out of such problems.

Appendix.

In what follows, we show that the matrix $B_{n,N}$ defined by (5) and (6) satisfies (15) and (22) for some constants $C_6, C_8, r' > 0$ with $1 < r' < \ell$ under the assumption 2 in Definition 2. The matrix B_n defined by (4) exists if (22) holds.

Proof of (15). Since

$$\| \ell^{-1}A_k \| \leq \ell^{-1} \| A \| + \ell^{-1} \| A - A_k \| \\ < 1 + \ell^{-1}C_2\delta^{-k},$$

we have

$$\| B_{n,N} \| < \prod_{k=n+1}^{N} (1 + \ell^{-1}C_2\delta^{-k}) < C_6. \quad \blacksquare$$

Proof of (22). Since

$$B_{n_1,n_2} - (\ell^{-1}A)^{n_2-n_1} = \sum_{k=n_1}^{n_2-1} B_{n_1,k}\ell^{-1}(A_{k+1} - A)(\ell^{-1}A)^{n_2-k-1},$$

we have

$$\| B_{n_1,n_2} - (\ell^{-1}A)^{n_2-n_1} \| < \sum_{k=n_1}^{n_2-1} C_6\ell^{-1}C_2\delta^{-k-1} \\ < C_9\delta^{-n_1}.$$

Then

$$\| B_{k,N} - B_{k,N'} \| \\ \leq \| B_{k,N} - (\ell^{-1}A)^{N-k} \| + \| B_{k,N'} - (\ell^{-1}A)^{N'-k} \| \\ + \| (\ell^{-1}A)^{N-k}(1 - (\ell^{-1}A)^{N'-N}) \| \\ < 2C_9\delta^{-k} + C_{10}(\frac{\ell_1}{\ell})^{N-k},$$

where $\ell_1 < \ell$ is a positive constant such that the absolute values of eigenvalues of the matrix A except ℓ are less than ℓ_1. Therefore we have, for $m \leq k \leq N$,

$$\| B_{m,N} - B_{m,N'} \| \leq \| B_{m,k} \| \| B_{k,N} - B_{k,N'} \| \\ < C_6(2C_9\delta^{-k} + C_{10}(\frac{\ell_1}{\ell})^{N-k}).$$

We here assume that $N - m$ is even without loss of generality and put $k = (N + m)/2$. Then the above estimate turns out to be

$$\| B_{m,N} - B_{m,N'} \| < C_6(2C_9\delta^{-\frac{N+m}{2}} + C_{10}(\frac{\ell_1}{\ell})^{\frac{N-m}{2}}).$$

Choosing $r' = \min(\delta^{1/2}, (\ell/\ell_1)^{1/2})$, we obtain (22). Obviously, $1 < r' < \ell$ holds. ∎

REFERENCES

[1] M. T. Barlow, E. A. Perkins, *Brownian Motion on the Sierpinski Gasket,* Probab. Th. Rel. Fields **79** (1988) 543-623.

[2] B. A. Cevaschanov, *Branching Processes,* Nauka, Moscow, 1971.

[3] J. C. D'Souza, J .D. Biggins, *The Supercritical Galton-Watson Process in Varying Environments on one-type varying environment Galton-Watson process,* Stoc. Proc. Appl. **42** (1992).

[4] S. Goldstein, *Random walks and diffusion on fractals,* IMA Math. Appl. **8** (1987) 121-129.

[5] T. Hattori, *Construction of asymptotically one-dimensional continuous Markov process on Sierpinski gasket,* RIMS Kokyuroku **783** (1992) 46-64 *(in Japanese).*

[6] K. Hattori, T. Hattori, H. Watanabe, *Gaussian field theories on general networks and the spectral dimensions,* Progr. Theor. Phys. Supplement **92** (1987) 108-143.

[7] T. Kumagai, *Estimates of the transition densities for Brownian motion on nested fractals,* To appear in Probab. Th. Rel. Fields.

[8] S. Kusuoka, *A diffusion process on a fractal,* Proc. Taniguchi Symp. (1987) 251-274.

[9] T. Lindstrøm, *Brownian motion on nested fractals,* Mem. Amer. Math. Soc. **420** (1990) 1-128.

Tetsuya Hattori
Department of Information Sciences
Faculty of Engineering
Utsunomiya University
Ishii-cho
Utsunomiya 321
JAPAN

Hiroshi Watanabe
Department of Mathematics
Nippon Medical School
Nakahara-ku
Kawasaki 211
JAPAN

On the Covering Time of a Disc by Simple Random Walk in Two Dimensions

GREGORY F. LAWLER

1. Introduction

Let $S(j)$ denote simple random walk taking values in Z^2. It is well known that $S(j)$ is recurrent and hence every finite set is eventually covered by the path of the walk. Let B_n be the discrete ball of radius n,

$$B_n = \{z : |z|^2 < n\},$$

and let T_n be the covering time of B_n,

$$T_n = \inf\{j : B_n \subset S[0,j]\},$$

where $S[0,j] = \{S(i) : 0 \leq i \leq j\}$. We will consider the limiting distribution of T_n. This random variable is closely related to the random variable

$$W_m = \sup\{n : B_n \subset S[0,m]\}.$$

Kesten [unpublished] and Révész [2, 3] independently have considered T_n and have proved that there exist $0 < \alpha_1 < \alpha_2 < \infty$ such that for all $t > 0$,

$$\begin{aligned} e^{-\alpha_2/t} &\leq \liminf_{n\to\infty} P\{\ln T_n \leq t(\ln n)^2\} \\ &\leq \limsup_{n\to\infty} P\{\ln T_n \leq t(\ln n)^2\} \leq e^{-\alpha_1/t}, \end{aligned} \tag{1}$$

or equivalently,

$$\begin{aligned} e^{-\alpha_2 s} &\leq \liminf_{m\to\infty} P\{\frac{[\ln W_m]^2}{\ln m} \geq s\} \\ &\leq \limsup_{m\to\infty} P\{\frac{[\ln W_m]^2}{\ln m} \geq s\} \leq e^{-\alpha_1 s}. \end{aligned}$$

Révész has verified this with $\alpha_2 = 120, \alpha_1 = 1/4$.

Kesten and Révész both made the natural conjecture: there exists an $\alpha > 0$ such that

$$\lim_{n\to\infty} P\{\ln T_n \leq t(\ln n)^2\} = e^{-\alpha/t}. \tag{2}$$

Kesten has also made a conjecture that this is true for $\alpha = 4$. His intuitive argument, which we sketch below, can be used to prove that (1) holds for $\alpha_2 = 4$. We are unable to prove the conjecture (2); however, in this paper we do improve the bounds on α_1 by proving the following.

Theorem 1.1. *If T_n is defined as above, then*

$$\begin{aligned} e^{-4/t} &\leq \liminf_{n\to\infty} P\{\ln T_n \leq t(\ln n)^2\} \\ &\leq \limsup_{n\to\infty} P\{\ln T_n \leq t(\ln n)^2\} \leq e^{-2/t}. \end{aligned}$$

We first sketch the ideas behind Kesten's conjecture because a similar idea will be used in our proof. Break up the random walk $S[0,j]$ into separate excursions between the ball of radius $2n$ and the sphere of radius n^2. To be more precise, let ∂B_n denote the boundary of B_n,

$$\partial B_n = \{z \notin B_n : |z-y| = 1 \text{ for some } y \in B_n\},$$

and $\xi_n = \xi(n)$ the hitting time of ∂B_n,

$$\xi_n = \inf\{j \geq 1 : S(j) \in \partial B_n\}.$$

Define times $k(1,n) < s(1,n) < k(2,n) < s(2,n) < \cdots$ by $k(1,n) = 0, s(1,n) = \xi_{n^2}$, and for $i > 1$,

$$k(i,n) = \inf\{j > s(i-1,n) : S(j) \in \partial B_{2n}\},$$

$$s(i,n) = \inf\{j > k(i,n) : S(j) \in \partial B_{n^2}\}.$$

Let $\bar{A}(i,n)$ be the set of points in B_n covered by the ith excursion,

$$\bar{A}(i,n) = B_n \cap S[k(i,n), s(i,n)].$$

For fixed n, $\bar{A}(1,n), \bar{A}(2,n), \bar{A}(3,n), \ldots$ are almost independent, identically distributed random subsets of B_n taken from the distribution "choose a random point on ∂B_{2n}, start a random walk until it reaches ∂B_{n^2}, and take the points in B_n convered by the walk." (These sets are not exactly i.i.d., but one can show that the error is not large and that in fact we can consider them to be such.) The problem thus reduces to an i.i.d. covering problem.

Let $A(1,n), A(2,n), A(3,n), \ldots$ be independent, identically distributed random subsets of B_n obtained from the distribution: choose $x \in \partial B_{2n}$

according to harmonic measure (hitting probability from infinity); perform a random walk until the walk reaches ∂B_{n^2}; and take the points in B_n visited by the walk. We now ask the question: how many random sets do we need to completely cover B_n? Let

$$V(k,n) = \cup_{i=1}^{k} A(i,n).$$

For a given $x \in B_n$, random walk estimates give that $P\{x \in A(1,n)\} \approx 1/2$. Hence by using independence, if $u > 0$,

$$(3)\ P\{x \notin V(u \ln n, n)\} \approx [P\{x \notin A(i,n)\}]^{u \ln n} \approx (1/2)^{u \ln n} = n^{-u \ln 2}.$$

Hence the expected number of points in $B_n \setminus V(u \ln n, n)$ is of order $n^{2-u\ln 2}$. This gives an immediate upper bound on the number of sets needed—if $u > 2/\ln 2$,

$$\lim_{n\to\infty} P\{V(u \ln n, n) = B_n\} = 1.$$

However, there are very strong correlations between points in the ball and estimating the probability by the expectation is not a sharp estimate. It is tempting, however, at this point to make the conjecture that for some $u > 0$, and all $\epsilon > 0$,

$$(4) \qquad \lim_{n\to\infty} P\{V((u-\epsilon)\ln n, n) = B_n\} = 0,$$

$$(5) \qquad \lim_{n\to\infty} P\{V((u+\epsilon)\ln n, n) = B_n\} = 1.$$

Consider the fraction of points covered in one excursion,

$$Y_n = |A(1,n)|/|B_n|,$$

where $|\cdot|$ denotes cardinality. The moments of Y_n can be estimated fairly easily and it can be shown (we omit the proof) that as $n \to \infty$, Y_n converges in distribution to a uniform random variable on $[0,1]$. We are then led to consider the following covering problem. Let $D(1,n), D(2,n), \ldots$ be indepedent, identically distributed random subsets of B_n obtained from the following distribution: for each i, choose a number p_i from the uniform distribution on $[0,1]$; then take $D(i,n)$ to be the random subset of B_n obtained by choosing sites independently each with probability p_i. This becomes a standard type of covering problem. Essentially, the fraction of points not covered by $D(1,n) \cup \cdots \cup D(k,n)$ is given (approximately) by $U_1 U_2 \cdots U_k$, where $U_1, U_2, \ldots$ are independent, uniform on $[0,1]$. We continue until the fraction of points not covered is of order n^{-2} (i.e., the number of points is of order 1), or equivalently we continue until

$$\sum_{i=1}^{k} \ln U_i \approx -2\ln n.$$

By the strong law of large numbers, we choose $k \approx -2E(\ln U_1)\ln n = 2\ln n$. This can be made rigorous to show that for every $\epsilon > 0$,

$$\lim_{n\to\infty} P\{\cup_{i=1}^{(2-\epsilon)\ln n} D(i,n) = B_n\} = 0,$$

$$\lim_{n\to\infty} P\{\cup_{i=1}^{(2+\epsilon)\ln n} D(i,n) = B_n\} = 1.$$

This intuition leads to the conjecture that (4) and (5) hold for $u = 2$.

One can adapt the argument to show that $u = 2$ is an upper bound. Consider excursions between the ball of radius n and the ball of radius $n^{1+\delta}$ where $\delta > 0$. If we let $\tilde{V}(k,n)$ be the analogue of $V(k,n)$ above for excursions from $2n$ to $n^{1+\delta}$, random walk estimates again give this analogue to (3),

$$P\{x \notin \tilde{V}(\tilde{u}\ln n)\} \approx [1/(1+\delta)]^{\tilde{u}\ln n} = n^{-\tilde{u}\ln(1+\delta)}.$$

Then again by approximating the probability by the expectation, we can show for every $\tilde{u} > 2/\ln(1+\delta)$,

$$\lim_{n\to\infty} P\{\tilde{V}(\tilde{u}\ln n, n) = B_n\} = 1.$$

Now using the fact that there are on the average $1/\delta$ excursions between n and $n^{1+\delta}$ for each excursion between n and n^2 we can conclude that for every $u > 2\delta/\ln(1+\delta)$,

$$\lim_{n\to\infty} P\{V(u\ln n) = B_n\} = 1.$$

By choosing δ sufficiently close to 0 we can then show that the above holds for any $u > 2$. (This is only a rough sketch of this argument; however, since this upper bound will follow rigorously from the work in this paper we will not give any more details at this point.)

The discussion so far leads to two natural conjectures that we state now.

Conjecture 1. Let $\sigma(n) = \inf\{k : S[0, s(k,n)] = B_n\} = \inf\{k : T_n < s(k,n)\}$. Then there exists a $u > 0$ such that

$$\frac{\sigma(n)}{\ln n} \to u,$$

in probability.

Conjecture 2. $u = 2$.

If Conjecture 1 is true, we can prove (2). To see this, we start with an estimate [see Lemma 3.1] that if $x \in B_{n^2}$, $a > 0$

$$P^x\{\xi_n < \xi(\exp(a(\ln n)^2))\} = 1 - (a \ln n)^{-1} + O((\ln n)^{-2}),$$

and hence

$$\lim_{n\to\infty} P\{s(u \ln n, n) < \xi(\exp(a(\ln n)^2))\} = e^{-u/a}.$$

Since $\xi(\exp(a(\ln n)^2))$ is about $\exp(2a(\ln n)^2))$ (see (6) for a more precise statement), we can conclude that

$$\lim_{n\to\infty} P\{s(u \ln n, n) < \exp(t(\ln n)^2)\} = e^{-2u/t},$$

and hence

$$\lim_{n\to\infty} P\{\ln T_n < t(\ln n)^2\} = e^{-2u/t}.$$

Note that $u = 2$ corresponds to $\alpha = 4$ in (2).

2. Preliminary results

As seen above, by considering shorter excursions, sharper estimates of the covering time can be made. We will therefore try to make the excursions as short as possible. One restriction, however, is that the excursions must be sufficiently long so that the intersection of the different excursions with the ball of radius n are almost independent. It turns out that the best we can do is excursions from radius $2n$ to radius $n(\ln n)^\alpha$ for some α. The main technical tool for showing asymptotic independence is the following lemma. Let $H_n(x, y)$ denote the hitting probability of ∂B_n, i.e.,

$$H_n(x, y) = P^x\{S(\xi_n) = y\}.$$

Let $H_n(y)$ denote (discrete) harmonic measure (from infinity), i.e.,

$$H_n(y) = \lim_{|x|\to\infty} H_n(x, y).$$

Then the following lemma combines Theorem 1.7.1, Lemma 2.1.2, and Theorem 2.1.3 of [1].

Lemma 2.1. *There exists a constant $c < \infty$ such that for all $n > 0$ and all $\alpha > 1$, if $x \in \partial B_n$, $y \in \partial B_{n(\ln n)^\alpha}$,*

$$|H_{n(\ln n)^\alpha}(x, y) - H_{n(\ln n)^\alpha}(0, y)| \le c(\ln n)^{-\alpha} H_{n(\ln n)^\alpha}(0, y).$$

$$|H_n(y, x) - H_n(x)| \le c(\ln n)^{1-\alpha} H_n(x).$$

Moreover, if $A \subset B_n$ and

$$\tau = \tau_A = \inf\{j \ge 1 : S(j) \in A\},$$

then

$$|P^x\{S(\xi(n(\ln n)^\alpha)) = y \mid \xi(n(\ln n)^\alpha) < \tau\} - H_{n(\ln n)^\alpha}(0,y)| \leq c(\ln n)^{1-\alpha} H_{n(\ln n)^\alpha}(0,y),$$

and

$$|P^y\{S(\xi(n)) = x \mid \xi(n) < \xi(n(\ln n)^\alpha)\} - H_n(x)| \leq c(\ln n)^{1-\alpha} H_n(x).$$

We now will choose $\alpha = 3$. This is the smallest value of α so that we can use the above lemma to prove asymptotic independence. We now consider excursions from the sphere of radius $2n$ to the sphere of radius $n(\ln n)^3$. The following lemma is essentially a corollary of Lemma 2.1 so we will not give a proof.

Lemma 2.2. *For $A \subset B_n$, let $Q_n(A)$ be given by*

$$Q_n(A) = \sum_{x \in \partial B_{2n}} H_{2n}(x) P^x\{S[0, \xi(n(\ln n)^3)] \cap B_n = A\}.$$

If $y, z \in \partial B_{n(\ln n)^3}$, let $Q_n(A; y, z)$ equal

$$P^y\{S[0, \xi(n(\ln n)^3)] \cap B_n = A \mid S(\xi(n(\ln n)^3)) = z, \xi(2n) < \xi(n(\ln n)^3))\}.$$

Then there exists a $c < \infty$ such that for all $A \subset B_n$ and all $y, z \in \partial B_{n(\ln n)^3}$,

$$|Q_n(A) - Q_n(A; y, z)| \leq c(\ln n)^{-2} Q_n(A).$$

As before, we define times $k(1,n) < s(1,n) < k(2,n) < s(2,n) < \cdots$ by $k(1,n) = 0, s(1,n) = \xi_{n(\ln n)^3}$, and for $i > 1$,

$$k(i,n) = \inf\{j > s(i-1,n) : S(j) \in \partial B_{2n}\},$$

$$s(i,n) = \inf\{j > k(i,n) : S(j) \in \partial B_{n(\ln n)^3}\}.$$

Again, let $\bar{A}(i,n)$ be the set of points covered by the ith excursion,

$$\bar{A}(i,n) = B_n \cap S[k(i,n), s(i,n)].$$

By Lemma 2.1, if $x \in \partial B_{2n}$,

$$|P\{S(k(i+1,n)) = x \mid \bar{A}(1,n), \ldots, \bar{A}(i,n)\} - H_{2n}(x)| \leq c(\ln n)^{-2} H_{2n}(x).$$

This means that if we consider $u(\ln n)^2/\ln\ln n$ excursions (which is the number it will require to cover B_n), and instead of starting the ith at $S(k(i,n))$ we start it at a point chosen by harmonic measure H_{2n}, independent of the past, we will get a different set with probability at most $O(1/\ln\ln n)$. To be more precise, let $A(1,n), A(2,n), \ldots$ be independent, identically distributed random subsets of B_n each obtained from the following distribution: choose $x \in \partial B_{2n}$ according to harmonic measure, start a random walk at x until it reaches $\partial B_{n(\ln n)^3}$, then choose the points of B_n visited by the walk. As before, let

$$V(k,n) = \cup_{i=1}^{k} A(i,n).$$

Then the following lemma can easily be deduced from Lemma 2.2. For ease of notation, for the remainder of this paper we will let $\phi_n = (\ln n)^2 / \ln \ln n$.

Lemma 2.3. *For any $\Gamma \subset B_n$ and $u > 0$ let $\bar{Q}(\Gamma) = \bar{Q}_{u,n}(\Gamma)$ be defined by*

$$\bar{Q}(\Gamma) = P\{V([u\phi_n], n) = \Gamma\},$$

and define $\tilde{Q}(\Gamma) = \tilde{Q}_{u,n}(\Gamma)$, by

$$\tilde{Q}(\Gamma) = P\{S[k(2,n), s([u\phi_n]+1, n)] \cap B_n = \Gamma\}.$$

Then for every $u > 0$ there exists a $c < \infty$ such that

$$\sum_{\Gamma \subset B_n} |\bar{Q}(\Gamma) - \tilde{Q}(\Gamma)| \leq c(\ln \ln n)^{-1}.$$

The distribution of $S[0, s(1,n)] \cap B_n$, of course, is significantly different from that of $A(1,n)$. However, it is not very difficult (we omit the argument) to see that it contributes very little to the covering time. We can thus prove the following lemma.

Lemma 2.4. *For any $u > 0$,*

$$\lim_{n\to\infty} P\{V(u\phi_n, n) = B_n\} = \lim_{n\to\infty} P\{S[0, s(u\phi_n, n)] \cap B_n = B_n\},$$

provided that at least one of the limits exists and equals 0 or 1.

For any $x \in \partial B_{n(\ln n)^3}$ and any $a > 0$,

$$\lim_{n\to\infty} (a\phi_n/3) P^x\{\xi(2n) \geq \xi(\exp\{a(\ln n)^2\})\} = 1$$

(see Lemma 3.1) . Hence

$$\lim_{n\to\infty} P\{s(u\phi_n, n) \leq \xi(\exp\{a(\ln n)^2\})\} = e^{-3u/a}.$$

For every $\epsilon > 0$, with probability one for all sufficiently large n,

$$\exp((2-\epsilon)a(\ln n)^2) \leq \xi(\exp(a(\ln n)^2)) \leq \exp((2+\epsilon)a(\ln n)^2) \tag{6}$$

(see [1, p. 146]). Hence

$$\lim_{n\to\infty} P\{s(u\phi_n, n) \le \exp(t(\ln n)^2)\} = e^{-6u/t}. \tag{7}$$

Suppose for some $u > 0$,

$$\lim_{n\to\infty} P\{s(u\phi_n, n) \le T_n\} = 1.$$

Then it follows from (7) that

$$\limsup_{n\to\infty} P\{\ln T_n \le t(\ln n)^2\} \le e^{-6u/t}.$$

Similarly,

$$\lim_{n\to\infty} P\{s(u\phi_n, n) \le T_n\} = 0,$$

implies

$$\liminf_{n\to\infty} P\{\ln T_n \le t(\ln n)^2\} \ge e^{-6u/t}.$$

Theorem 1.1 then follows from Lemma 2.4 and the following theorem.

Theorem 2.5. *For every $u < 1/3$,*

$$\lim_{n\to\infty} P\{V(u\phi_n, n) = B_n\} = 0.$$

For every $u > 2/3$,

$$\lim_{n\to\infty} P\{V(u\phi_n, n) = B_n\} = 1.$$

As a preliminary to proving this theorem, we will prove the following proposition which states roughly that if there is some chance that the ball is not covered in $u\phi_n$ excursions, then there is no chance that it is covered in $u(1-\epsilon)\phi_n$ excursions.

Proposition 2.6. *Suppose for some $0 < u, c, \alpha < \infty$ and all n*

$$P\{V(u\phi_n, n) \ne B_n\} \ge c(\ln n)^{-\alpha}.$$

Then for all $v < u$,

$$\lim_{n\to\infty} P\{V(v\phi_n, n) = B_n\} = 0.$$

One corollary of this proposition is the following: Suppose for every u the limit

$$\lim_{n\to\infty} P\{V(u\phi_n, n) = B_n\} = f(u)$$

exists. Then there must be a unique u_0 such that $f(u) = 0$ for $u < u_0$ and $f(u) = 1$ for $u > u_0$. To say this differently, let

$$\sigma(n) = \inf\{k : B_n \subset S[0, s(k,n)]\}.$$

Then if $\sigma(n)/\phi_n$ converges in probability to a random variable, the limit must be a constant.

Proof of Proposition 2.6. Fix α as in the hypotheses. For any x, we let

$$B_m(x) = \{z : |z - x| < m\},$$

and $\partial B_m(x)$ similarly. Let $r = r_n = n(\ln n)^{-\alpha-5}$ and $J = J_n = [(\ln n)^{\alpha+1}]$. Find points $x_1, \ldots, x_J$ such that $B_{r(\ln r)^3}(x_1), \ldots, B_{r(\ln r)^3}(x_J)$ are disjoint subsets of B_n (this can be done for n sufficiently large). Let $R_i = \partial B_{r(\ln r)^3}(x_i)$, $W_i = \partial B_{2r}(x_i)$.

Choose $v < u$, set $K = K_n = [v\phi_n] + 1$, and consider the random sets $A(1,n), \ldots, A(K,n)$, obtained from independent random walks, $S^1, \ldots, S^K$ starting with harmonic measure on ∂B_{2n} stopped upon reaching $\partial B_{n(\ln n)^3}$. For a given i we will consider the number of excursions to the set $B_{2r}(x_i)$. We will say that such an excursion occurs between times $s < t$ for the kth walk if: $S^k(s) \in R_i, S^k(t) \in R_i; S^k(j) \notin R_i, s < j < t$; and $S^k[s,t] \cap W_i \neq \emptyset$. A given random walk may have more than one such excursion for a given i. Let U_i be the total number of excursions to $B_{2r}(x_i)$ by the random walks $S^1, \ldots, S^K$. Let v_1, v_2 satisfy $v < v_1 < v_2 < u$ and suppose that $U_i \le v_1\phi_n$. Then (if n is sufficiently large) $U_i \le v_2\phi_r$. By Lemma 2.2, we then get

$$P\{B_r(x_i) \not\subset V(K,n)\} \ge (1/2)P\{B_r \not\subset V(v_2\phi_r, r)\} \ge c_1(\ln n)^{-\alpha}.$$

Lemma 2.2 actually says much more. The conditional probability that $B_r(x_i) \not\subset V(K,n)$ given the values of $S^1, \ldots, S^K$ outside the excursions to $B_{2r}(x_i)$ (that is, given $S^k(j)$ for all values of j except for $s < j < t$ if S^k has an excursion to $B_r(x_i)$ between s and t) is bounded below by $c_1(\ln n)^{-\alpha}$. Because of this "conditional independence" we can show that if Y_n is the number of indices i between 1 and K such that $U_i \le v_1\phi_n$, then

$$\begin{aligned} P\{B_n \subset V(v\phi_n, n)\} &\le P\{B_r(x_1) \cup \cdots \cup B_r(x_J) \subset V(v\phi_n, n)\} \\ &\le (1 - c_1(\ln n)^{-\alpha})^{Y_n}. \end{aligned}$$

Hence in order to prove the proposition it suffices to prove that for some $v_1 < u, c_2 > 0$,

$$P\{U_i \le v_1 \phi_n \text{ for at least } c_2(\ln n)^{\alpha+1} \text{ values of } i\} \to 1.$$

This becomes a straightforward estimate of the mean and variance of U_i. Note that for fixed i, U_i is a sum of i.i.d. random variables

$$U_i = X_1 + \cdots + X_K,$$

where X_k denotes the number of excursions to $B_r(x_i)$ by the kth random walk. If $x \in \partial B_{2n}$, and $\tau = \tau(i,n) = \inf\{j : S^k(j) \in W_i\}$, then (see Lemma 3.1),

$$P^x\{X_k > 0\} = P^x\{\tau < \xi(n(\ln n)^3)\} = 3/(\alpha+8) + o(1),$$

where the $o(1)$ term is bounded uniformly in x and i. Similarly if $y \in R_i$,

$$P^y\{\tau < \xi(n(\ln n)^3))\} = (\alpha+5)/(\alpha+8) + o(1),$$

and hence for $m \ge 1$,

$$P^x\{X_k \ge m+1 \mid X_k \ge m\} = (\alpha+5)/(\alpha+8) + O((\ln n)^{-1}).$$

Therefore, $E(X_k) = 1 + o(1)$, $\text{Var}(X_k) = O(1)$, and

$$E(U_i) = v\phi_n + o(\phi_n), \quad Var(U_i) = O(\phi_n).$$

By Chebyshev's inequality,

$$P\{U_i \ge v_1\phi_n\} \le O(\phi_n^{-1}).$$

If we now let Z_i denote the indicator function of the event $\{U_i \ge v_1\phi_n\}$, we have

$$E(\sum_{i=1}^{J} Z_i) \le O(\phi_n^{-1}(\ln n)^{\alpha+1}),$$

and hence by Markov's inequality,

$$P\{\sum_{i=1}^{J} Z_i \ge J/2\} = O(\phi_n^{-1}).$$

This completes the proof. *Q.E.D.*

3. Proof of Theorem

Let $A(k,n)$ and $V(k,n)$ be defined as in Section 2. For every u let $R(u,n) = B_n \setminus V(u\phi_n, n)$ be the set of points not visited after $u\phi_n$ excursions and let $Z(u,n)$ be the cardinality of $R(u,n)$. Essentially, the proof requires only estimating the first two moments of $Z(u,n)$. We will show that

$$E(Z(u,n)) \approx n^{4-6u},$$

where $\approx$ means that the logarithms of both sides are asymptotic. This will clearly show that for $u > 2/3$,

$$\lim_{n\to\infty} P\{Z(u,n) > 0\} \le \lim_{n\to\infty} E(Z(u,n)) = 0.$$

For $u < 1/3$, we will show that

$$E(Z(u,n)^2) \le C(\ln n)^r [E(Z(u,n))]^2,$$

for some $r = r(u)$ from which it can be deduced that

$$P\{Z(u,n) > 0\} \ge (C \ln n)^{-r}.$$

The theorem will then follow from Proposition 2.6. It will also follow from our calculations, that for any $u > 1/3$, there exist $c, \epsilon > 0$ such that

$$E(Z(u,n)^2) \ge cn^{\epsilon}[E(Z(u,n)]^2,$$

so this argument cannot be used to analyze the behavior for $u \in (1/3, 2/3)$. We start with two lemmas about simple random walk.

Lemma 3.1. [1, Exercise 1.6.8] *if* $n < m < N$ *and* $x \in \partial B_m$,

$$P^x\{\xi(n) < \xi(N)\} = \frac{\ln N - \ln m + O(n^{-1})}{\ln N - \ln n},$$

where the $O(\cdot)$ *term is bounded uniformly in* x, m, N.

Lemma 3.2. [1, Proposition 1.6.7] *There exists a constant* γ *such that if* $x \in B_n$, $|x| \ge \ln n$, *then*

$$P^x\{0 \in S[0, \xi(n)]\} = \frac{\ln n - \ln |x|}{\ln n + \gamma} + O((\ln n)^{-2}),$$

where the $O(\cdot)$ *term is bounded uniformly in* x.

From this we get an immediate corollary.

Corollary 3.3. *For any $x, y \in B_{3n}$, $|x-y| \geq \ln n$,*

$$P^x\{y \in S[0, \xi(n(\ln n)^3)]\} =$$

$$3\frac{\ln\ln n}{\ln n} - \frac{\ln a}{\ln n} + 3\ln a\frac{\ln\ln n}{(\ln n)^2} + \gamma\ln a\frac{1}{(\ln n)^2} + O((\frac{\ln\ln n}{\ln n})^2),$$

where $a = |x-y|/n$ and the $O(\cdot)$ term is bounded uniformly in x, y.

Proof. Note that

$$\begin{aligned} P^{x-y}\{0 \in S[0, \xi(n(\ln n)^3 - 6n)]\} &\leq P^x\{y \in S[0, \xi(n(\ln n)^3]\} \\ &\leq P^{x-y}\{0 \in S[0, \xi(n(\ln n)^3 + 6n)]\}, \end{aligned}$$

and hence the result follows from Lemma 3.2. *Q.E.D.*

Lemma 3.4. *If $H_{2n}(z)$ denotes harmonic measure on ∂B_{2n}, $x \in B_n$,*

$$\sum_{z \in \partial B_{2n}} H_{2n}(z) \ln|x-z| = \ln n + \ln 2 + O((\ln n)^{-1}),$$

where the $O(\cdot)$ term is uniform in x.

Proof. If $|x| < 3\ln n$, then for every $z \in \partial B_{2n}$,

$$\ln|x-z| = \ln n + \ln 2 + O(\ln n/n),$$

so without loss of generality we may assume $|x| \geq 3\ln n$ (all $O(\cdot)$ terms in this proof will be bounded uniformly in points x, y, z). Let $\tau_x = \inf\{j : S(j) = x\}$. Take $y \in \partial B_{n^3}$. Then by Lemma 3.2 (and the trick in the proof of Corollary 3.3),

$$P^y\{\tau_x < \xi(n^3)\} = \frac{1}{3} - \frac{\gamma}{9\ln n} + O((\ln n)^{-2}).$$

By Lemma 3.1,

$$P^y\{\xi(2n) < \xi(n^3)\} = \frac{1}{2} + \frac{\ln 2}{4\ln n} + O((\ln n)^{-2}).$$

Note that

$$P^y\{\tau_x < \xi(n^3)\} = P^y\{\xi(2n) < \xi(n^3)\}P^y\{\tau_x < \xi(n^3) \mid \xi(2n) < \xi(n^3)\},$$

and hence

$$P^y\{\tau_x < \xi(n^3) \mid \xi(2n) < \xi(n^3)\} = \frac{2}{3} - \frac{\ln 2}{3\ln n} - \frac{2\gamma}{9\ln n} + O((\ln n)^{-2}).$$

Let $h(z) = h_n(z, y)$ be the conditional measure on ∂B_{2n} given $\xi(2n) \leq \xi(n^3)$,

$$h(z) = P^y\{S(\xi(2n)) = z \mid \xi(2n) \leq \xi(n^3)\}.$$

Then, using Lemma 3.2,

$$\begin{aligned} P^y\{\tau_x \le \xi(n^3) \mid \xi(2n) \le \xi(n^3)\} \\ = \sum_{z \in \partial B_{2n}} h(z) P^z\{\tau_x \le \xi(n^3)\} \\ = \sum_{z \in \partial B_{2n}} h(z)[1 - \frac{(2/3)\gamma + \ln|z-x|}{3\ln n} + O((\ln n)^{-2})]. \end{aligned}$$

Hence,

$$\sum_{z \in \partial B_{2n}} h(z) \ln|z-x| = \ln n + \ln 2 + O((\ln n)^{-1}).$$

We now can use Lemma 2.1 to conclude that

$$|h(z) - H_{2n}(z)| \le O((\ln n)^{-2}),$$

(we can actually do much better than this) and hence

$$\sum_{z \in \partial B_{2n}} H_{2n}(z) \ln|x-z| = \ln n + \ln 2 + O((\ln n)^{-1}). \quad Q.E.D.$$

Corollary 3.5. *If $x \in B_n$,*

$$P\{x \in A(k,n)\} = 3\frac{\ln\ln n}{\ln n} - \frac{\ln 2}{\ln n} + O((\frac{\ln\ln n}{\ln n})^2).$$

where the $O(\cdot)$ term is bounded uniformly for $x \in B_n$.

Lemma 3.6. *For every $u > 0$, there exist constants $c, \beta < \infty$ such that for all $x \in B_n$,*

$$\begin{aligned} c^{-1}(\ln n)^{-\beta} n^{-3u} \exp\{\frac{u \ln 2 \ln n}{\ln\ln n}\} &\le P\{x \in R(u,n)\} \\ &\le c(\ln n)^{\beta} n^{-3u} \exp\{\frac{u \ln 2 \ln n}{\ln\ln n}\}. \end{aligned}$$

Proof. By Corollary 3.5,

$$\begin{aligned} P\{x \in R(u,n)\} &= [1 - 3\frac{\ln\ln n}{\ln n} + \frac{\ln 2}{\ln n} + O((\frac{\ln\ln n}{\ln n})^2))]^{u\phi_n} \\ &= n^{-3u} \exp\{\frac{u \ln 2 \ln n}{\ln\ln n}\}(1 + O((\frac{\ln\ln n}{\ln n})^2))^{u\phi_n}. \quad Q.E.D. \end{aligned}$$

It follows immediately that

$$(8) \quad c^{-1}(\ln n)^{-\beta} \le E(Z(u,n))[n^{2-3u} \exp\{\frac{u \ln 2 \ln n}{\ln\ln n}\}]^{-1} \le c(\ln n)^{\beta},$$

where c, β depend on u. If $u > 2/3$, then it follows that

$$\lim_{n\to\infty} P\{V(u\phi_n, n) \neq B_n\} \leq \lim_{n\to\infty} E(Z(u,n)) = 0.$$

To get bounds in the other direction, we will consider the second moment of $Z(u,n)$, which is given by

$$E(Z(u,n)^2) = \sum_{x,y\in B_n} P\{x,y \in R(u,n)\}.$$

Lemma 3.7. *Suppose $x, y \in B_n$ with $|x-y| \geq \ln n$. Let $b = (\ln n - \ln|x-y|)/\ln n$. Then*

$$P\{x,y \notin A(1,n)\} = 1 - (\frac{1}{1+b})[6\frac{\ln\ln n}{\ln n} - \frac{2\ln 2}{\ln n}] + O((\frac{\ln\ln n}{\ln n})^2),$$

where the $O(\cdot)$ term is bounded uniformly in x, y.

Proof. Suppose $x, y \in B_n$. Let I_x be the indicator function of the event $\{x \in A(1,n)\}$ and similarly for I_y. Then by Corollary 3.5,

$$E(I_x + I_y) = 6\frac{\ln\ln n}{\ln n} - \frac{2\ln 2}{\ln n} + O((\frac{\ln\ln n}{\ln n})^2).$$

Also,

$$\begin{aligned} E(I_x + I_y) &= P\{I_x + I_y \geq 1\} + P\{I_x + I_y = 2\} \\ &= P\{I_x + I_y \geq 1\}(1 + P\{I_x + I_y = 2 \mid I_x + I_y \geq 1\}). \end{aligned}$$

By the strong Markov property and Corollary 3.3,

$$P\{I_x + I_y = 2 \mid I_x + I_y \geq 1\} = b + O(\frac{\ln\ln n}{\ln n}),$$

where $b = -\ln a/\ln n = (\ln n - \ln|x-y|)/\ln n$. The result follows from dividing the two terms.

The next lemma is then proved in the same way as Lemma 3.6.

Lemma 3.8. *For every u, there exist c_2, ν such that if $x, y \in B_n$ with $|x-y| \geq \ln n$ and $b = (\ln n - \ln|x-y|)/\ln n$,*

$$\begin{aligned} & c_2^{-1}(\ln n)^{-\nu} n^{-6u/(1+b)} \exp\{\frac{2u\ln 2\ln n}{(1+b)\ln\ln n}\} \\ \leq\ & P\{x,y \in R(u,n)\} \\ \leq\ & c_2(\ln n)^{\nu} n^{-6u/(1+b)} \exp\{\frac{2u\ln 2\ln n}{(1+b)\ln\ln n}\}. \end{aligned}$$

With this lemma, we are now able to give an estimate for the second moment of $Z(u,n)$ for $u < 1/3$.

Lemma 3.9. *For each $u < 1/3$ there exist $C, \nu < \infty$ such that*

$$E((Z(u,n)^2) \leq C(\ln n)^\nu n^{4-6u} \exp\{\frac{2\ln 2 \ln n}{\ln\ln n}\}. \tag{9}$$

Proof. Throughout this proof we use C to denote a constant that may depend on u (but not on x or y) whose value may change from line to line. As noted before,

$$E(Z(u,n)^2) = \sum_{x,y\in B_n} P\{x,y \in R(u,n)\}.$$

If $|x-y| < \ln n$, we estimate by Lemma 3.6,

$$P\{x,y\in R(u,n)\} \leq P\{x \in R(u,n)\} \leq Cn^{-3u}(\ln n)^\beta \exp\{\frac{\ln 2\ln n}{\ln\ln n}\},$$

and hence the sum over all $|x-y| < \ln n$ is bounded by a constant times

$$n^2(\ln n)^{\beta+2} n^{-3u} \exp\{\frac{u\ln 2\ln n}{\ln\ln n}\},$$

which is clearly less than the right hand side of (9).

Let b be defined as in Lemma 3.8. Note that $b \geq -2(\ln n)^{-1}$ for $x, y \in B_n$. Hence by the lemma, if $|x-y| \geq \ln n$,

$$P\{x,y\in R(u,n)\} \leq C(\ln n)^\nu n^{-6u/(1+b)} \exp\{\frac{2u\ln 2\ln n}{\ln\ln n}\},$$

It therefore suffices to prove that

$$\sum_{|x-y|\geq \ln n} n^{-6u/(1+b)} \leq Cn^{4-6u},$$

or hence to prove

$$\sum_{|x|\leq n} n^{-6u/(1+b)} \leq Cn^{2-6u},$$

where here $b = (\ln n - \ln|x|)/\ln n$. To prove this estimate, let

$$D_j = \{2^{-j}n < |x| \leq 2^{-j+1}n\}, \quad j = 1,2,\ldots.$$

Then,

$$\begin{aligned}\sum_{x\in D_j} n^{-6u/(1+b)} &\leq C2^{-2j}n^2\exp\{-\frac{6u\ln n}{1+j\ln 2(\ln n)^{-1}}\}\\ &= Cn^2\exp\{-2j\ln 2-\frac{6u\ln n}{1+j\ln 2(\ln n)^{-1}}\}.\end{aligned}$$

Therefore,

$$\sum_{|x|\leq n} n^{-6u/(1+b)} \leq Cn^2\sum_{j=1}^{\infty} a_j,$$

where

$$a_j = a_{j,n} = \exp\{-2j\ln 2 - \frac{6u\ln n}{1+j\ln 2(\ln n)^{-1}}\}.$$

Note that

$$a_j/a_{j-1} \leq \exp\{-2\ln 2 + 6u\ln 2\}.$$

Hence for every $u < 1/3$, there exists $\rho < 1$ such that

$$a_j \leq \rho^{j-1}a_1,$$

and

$$\sum_{|x|\leq n} n^{-6u(1+b)} \leq Cn^2 a_1 \leq Cn^{2-6u}. \quad Q.E.D.$$

It follows immediately from this lemma and Lemma 3.6, that for $u < 1/3$,

$$E(Z(u,n)^2) \leq C(u)[E(Z(u,n))]^2(\ln n)^{\nu+2\beta}. \tag{10}$$

We now note that we cannot get such an estimate for $u > 1/3$. For any u, by Lemma 3.6,

$$[E(Z(u,n))]^2 \approx n^{4-6u}$$

(here we are using $\approx$ to mean that the logarithms of both sides are asymptotic). For any $\rho < 1$ there are on the order of $n^{2+2\rho}$ points with $|x-y|$ of order n^ρ. For such x, y, by Lemma 3.8,

$$P\{x, y \in R(u,n)\} \approx n^{-6u/(2-\rho)},$$

and hence

$$E(Z(u,n)^2) \geq \sum_{|x-y|\approx n^\rho} P\{x,y\in R(u,n)\} \approx n^{2+2\rho-(6u/(1+\rho))}.$$

For any $u > 1/3$, we can find $\rho < 1$ such that the above exponent is greater than $4-6u$ and hence we can find constants $c, \epsilon > 0$ such that

$$E(Z(u,n)^2) \geq cn^\epsilon[E(Z(u,n))]^2.$$

It is easy to finish the proof of Theorem 2.5. Take $u < 1/3$ and find let $v \in (u, 1/3)$. By (10),

$$E(Z(v,n)^2) \leq C(\ln n)^{\nu} E(Z(v,n)).$$

If X is any nonnegative random variable, a simple Jensen's inequality argument (see, e.g. [1, (3.5)]), gives that

$$P\{X > 0\} \geq [E(X)]^2 / E(X^2).$$

Hence $P\{Z(v,n) > 0\} \geq (C \ln n)^{-(\nu+2\beta)}$, i.e.,

$$P\{V(v\phi_n, n) \neq B_n\} \geq (C \ln n)^{-(\nu+2\beta)}.$$

We now use Proposition 2.6 to conclude

$$\lim_{n\to\infty} P\{V(u\phi_n, n) = B_n\} = 0.$$

4. Random walk on a torus

There is a similar problem concerning the covering time of a random walk on a discrete torus. Let

$$\Lambda_n = \{(z_1, z_2) \in Z^2 : -n \leq |z_i| \leq n\}$$

be a square centered at 0 with $(2n+1)^2$ points and give Λ_n periodic boundary conditions. Let $S(j)$ be simple random walk on Λ (which can be considered as a finite Markov chain) starting at 0 and

$$T = T_n = \inf\{j : S[0,j] = \Lambda_n\}$$

the covering time of the square. It has been shown that there exists $0 < c_1 < c_2 < \infty$ such that

$$\lim_{n\to\infty} P\{T \leq c_1 n^2 (\ln n)^2\} = 0,$$

$$\lim_{n\to\infty} P\{T \leq c_2 n^2 (\ln n)^2\} = 1.$$

Here we sketch how the ideas of this paper can be used to improve estimates on the lower bound c_1. In particular we show this holds for all $c_1 < 8/\pi$. We note that the upper bound can be proven for all $c_2 > 16/\pi$; however, this upper bound can be proven by simpler means and will not be discussed here.

Let $m = m_n = n(\ln n)^{-3}$ and let

$$\tilde{T} = \inf\{j : S[0,j] \subset B_m\},$$

where again B_m is the discrete ball of radius m. Clearly, $T \geq \tilde{T}$. Let $k(1,n) = 0, s(1,n) = \xi_n$, and for $i > 0$,

$$k(i,n) = \inf\{j > s(i-1,n) : S(j) \in \partial B_{2m}\},$$

$$s(i,n) = \inf\{j > k(i,n) : S(j) \in \partial B_n\}.$$

Then in the same way as Theorem 2.5 is proved, we can show that for all $u < 1/3$,

$$\lim_{n\to\infty} P\{s(u\phi_n, n) > \tilde{T}\} = 0.$$

Below we will sketch a proof that

$$\frac{s(u\phi_n, n)}{n^2(\ln n)^2} \to \frac{24u}{\pi}, \tag{11}$$

in probability. From this we then get for any $a < 8/\pi$

$$\lim_{n\to\infty} P\{T \leq an^2(\ln n)^2\} \leq \lim_{n\to\infty} P\{\tilde{T} \leq an^2(\ln n)^2\} = 0.$$

To prove (11), consider

$$\tau = \tau_n = \inf\{j > \xi_n : S(j) \in \partial B_{2m}\}.$$

Then for $x \in B_{2m}$, $\tau = \xi_n + (\tau - \xi_n)$. By [1, (1.21)],

$$E^x(\xi_n) = n^2 + O((\ln n)^{-6})$$

(here and below the $O(\cdot)$ terms are uniformly bounded for $x \in B_{2m}$.) Also, by Lemma 2.1 and the strong Markov property, if $x, y \in B_{2m}$,

$$E^x(\tau - \xi_n) = E^y(\tau - \xi_n)(1 + O((\ln n)^{-3})),$$

and hence

$$E^x(\tau) = E^y(\tau)(1 + O((\ln n)^{-3})).$$

This allows us to conclude that

$$E(s(u\phi_n, n)) = u\phi_n E^x(\tau)(1 + O((\ln n)^{-3})).$$

If we consider the number of visits to 0 before τ, then for $x \in B_{2m}$,

$$\begin{aligned} E^x \sum_{j=0}^{\tau} I\{S(j) = 0\} &= E^x \sum_{j=0}^{\xi_n} I\{S(j) = 0\} \\ &= G_n(x, 0) = (6/\pi)\ln\ln n + O(1), \end{aligned}$$

see [1, Proposition 1.6.7]. If $W(i,n)$ denotes the number of visits to 0 before $s(i,n)$, we get

$$E(W(u\phi_n, n)) = u\phi_n(6/\pi)(\ln\ln n + O(1)).$$

By standard Markov chain theory, the ratio $E(W(u\phi_n, n))/E(s(u\phi_n, n))$ must approach $(2n+1)^{-2}$, the invariant measure at 0. From this we conclude that

$$E^x(\tau) = 4n^2(6/\pi)\ln\ln n + O(n^2).$$

Therefore,

$$E(s(u\phi_n, n)) = (24u/\pi)n^2 \ln\ln n + O(n^2).$$

It is not too difficult to show that

$$\mathrm{Var}(s(u\phi_n, n)) = o(E(s(u\phi_n, n))^2),$$

and so the weak law (11) holds (we omit the details, but the key estimate is that $E^x(\tau^2) \le c[E^x(\tau)]^2$).

Acknowledgements

I would like to thank H. Kesten for useful discussions about this problem. This research was partially supported by the National Science Foundation.

References

1. G. Lawler (1991). *Intersections of Random Walks*. Birkhäuser-Boston.
2. P. Révész (1989). Simple symmetric random walk in R^d. Almost everywhere convergence. *Proc. Internat. Conf. on Almost Everywhere Convergence in Probab. and Ergodic Theory* (G. A. Edgar and L. Sucheston, eds.) Academic Press.
3. P. Révész (1990). Estimates on the largest disc covered by a random walk. Annals of Probab. **18**, 1784-1789.

Department of Mathematics
Duke University
Durham, NC 27706
jose@math.duke.edu

Φ-Variation of the Local Times of Symmetric Lévy Processes and Stationary Gaussian Processes

Michael B. Marcus and Jay Rosen

1 Introduction

Let $X = \{X(t), t \in \mathbf{R}^+\}$ be a symmetric real-valued Lévy process with characteristic function

$$Ee^{i\lambda X(t)} = e^{-t\psi(\lambda)} \tag{1.1}$$

and Lévy exponent

$$\psi(\lambda) = 2\int_0^\infty (1 - \cos u\lambda)\, d\nu(u) \tag{1.2}$$

for ν a Lévy measure, i.e. $\int_0^\infty (1 \wedge u^2)\, d\nu(u) < \infty$. We also include the case $\psi(\lambda) = \lambda^2/2$ which gives us standard Brownian motion.

In [4] we used the Dynkin Isomorphism Theorem to study the almost sure variation in the spatial variable of the local time of the symmetric stable processes of index $1 < \beta \leq 2$, i.e. $\psi(\lambda) = c|\lambda|^\beta$. In this note we will show how the proofs in [4] can be modified so as to generalize the almost sure variation results obtained there to a large class of symmetric Lévy processes with Lévy exponent *psi* regularly varying at infinity of order $1 < \beta \leq 2$.

Such Lévy processes X have an almost surely jointly continuous local time which we denote by $L = \{L_t^x, (t, x) \in \mathbf{R}^+ \times \mathbf{R}\}$, and normalize by requiring that

$$E^0\left(\int_0^\infty e^{-t}\, dL_t^x\right) = u^1(x)$$

where

$$u^1(x) = \frac{1}{\pi}\int_0^\infty \frac{\cos x\lambda}{1 + \psi(\lambda)}\, d\lambda$$

is the 1-potential density for X. We set

$$\sigma^2(x) = \frac{1}{\pi}\int_0^\infty \frac{1 - \cos x\lambda}{\psi(\lambda)}\, d\lambda. \tag{1.3}$$

It follows from Pitman [6] that if $\psi(\lambda)$ is regularly varying at infinity of order $1 < \beta \leq 2$, then $\sigma^2(x)$ is regularly varying at zero of order $\beta - 1$, and we have

$$\sigma^2(x) \sim c_\beta \frac{1}{x\psi(\frac{1}{x})} \qquad as\ x \to 0 \tag{1.4}$$

with c_β depending only on β. Throughout this paper we use the notation $f \sim g$ to mean that $\lim f/g = 1$.

Let $\pi = \{0 = x_0 < x_1 \cdots < x_{k_\pi} = a\}$ denote a partition of $[0, a]$, and let $m(\pi) = \sup_{1 \le i \le k_\pi}(x_i - x_{i-1})$ denote the length of the largest interval in π. $m(\pi)$ is called the mesh of π. Let $Q_a(\delta) = \{\text{partitions } \pi \text{ of } [0, a] \mid m(\pi) \le \delta\}$.

To clarify the notation in all that follows we note that in the expression $\sum_{x_i \in \pi} f(x_{i-1}, x_i)$, for some function f, we mean that the sum is taken over all the terms in which both x_{i-1} and x_i are contained in π.

Theorem 1 *Let $X = \{X(t), t \in \mathbf{R}^+\}$ be a real valued symmetric Lévy process with $\sigma^2(x)$ concave on $[0, \delta]$ and regularly varying at zero of order $\beta - 1$ where $1 < \beta \le 2$, and let $\{L_t^x, (t, x) \in \mathbf{R}^+ \times \mathbf{R}\}$ be the local time of X.*

i) Let $\Phi(x)$ denote any function which is an inverse for $\sigma(x)$ near $x = 0$. (Thus, $\Phi(x)$ is regularly varying at zero of order $2/(\beta - 1)$). If $\{\pi(n)\}$ is any sequence of partitions of $[0, a]$ such that $m(\pi(n)) = o(\Phi(\sqrt{1/\log n}))$ then

$$\lim_{n\to\infty} \sum_{x_i \in \pi(n)} \Phi(|L_t^{x_i} - L_t^{x_{i-1}}|) = c(\beta) \int_0^a |L_t^x|^{1/(\beta-1)}\, dx \tag{1.5}$$

for almost all $t \in \mathbf{R}^+$ almost surely, where

$$c(\beta) = \frac{2^{2/(\beta-1)}}{\sqrt{\pi}} \Gamma\left(\frac{1}{\beta-1} + \frac{1}{2}\right) \left(\frac{1}{\Gamma(\beta)\sin(\frac{\pi}{2}(\beta-1))}\right)^{1/(\beta-1)} \tag{1.6}$$

ii) Let $\Upsilon(x)$ denote any function which is an inverse for $\sigma(x)\sqrt{2\log\log 1/x}$ near zero. (Thus, $\Upsilon(x)$ is regularly varying at zero of order $2/(\beta - 1)$). Then

$$\lim_{\delta\to 0} \sup_{\pi \in Q_a(\delta)} \sum_{x_i \in \pi} \Upsilon(|L_t^{x_i} - L_t^{x_{i-1}}|) = c'(\beta) \int_0^a |L_t^x|^{1/(\beta-1)}\, dx \tag{1.7}$$

almost surely for each $t \in \mathbf{R}^+$, where

$$c'(\beta) = \left(\frac{1}{\Gamma(\beta)\sin(\frac{\pi}{2}(\beta-1))}\right)^{1/(\beta-1)}. \tag{1.8}$$

To prove Theorem 1 we use a corollary of the Dynkin Isomorphism Theorem, Lemma 4.3 of [5], which enables us to obtain almost sure results for variations of the local times of symmetric Lévy processes from analogous results about the variation of their associated Gaussian processes. The mean zero Gaussian process $\{G(x), x \in \mathbf{R}\}$ with covariance $g(x, y)$ is said to be associated with the Markov process X if $g(x,y) = u^1(x,y)$, the 1–potential density of X.

In particular, part ii) of Theorem 1 follows immediately by applying the methods of [4] to the results on Gaussian processes of Kawada and Kono [2]. Part i) of Theorem 1 follows from the next theorem on Gaussian processes in the same way that Theorem 1.1 followed from Theorem 1.2 in [4].

Theorem 2 *Let $\{G(x), x \in \mathbf{R}\}$ be a mean zero Gaussian process with stationary increments, such that*

$$\sigma^2(x) = E(G(x) - G(0))^2. \tag{1.9}$$

is concave and regularly varying at $x = 0$ of order $\beta - 1$ where $1 < \beta \leq 2$. Let $\Phi(x)$ denote any function which is an inverse for $\sigma(x)$ near $x = 0$. (Thus, $\Phi(x)$ is regularly varying at zero of order $2/(\beta - 1)$). If $\{\pi(n)\}$ is any sequence of partitions of $[0, a]$ such that $m(\pi(n)) = o(\Phi(\sqrt{1/\log n}))$ then

$$\lim_{n\to\infty} \sum_{x_i \in \pi(n)} \Phi(|G(x_i) - G(x_{i-1})|) = E|\eta|^{2/(\beta-1)} \qquad a.s. \tag{1.10}$$

where η is a normal random variable with mean 0 and variance 1. Also,

$$\lim_{n\to\infty} \sum_{x_i \in \pi(n)} \Phi(|G^2(x_i) - G^2(x_{i-1})|) \tag{1.11}$$

$$= E|\eta|^{2/(\beta-1)} 2^{2/(\beta-1)} \int_0^a |G(x)|^{2/(\beta-1)}\, dx$$

almost surely.

As in [4] the almost sure result of Theorem 1 part i) will lead to the following L^r convergence of the variation. Note that in Theorem 3 we do not require any conditions on the rate of convergence of $m(\pi(n))$ to zero. In addition, our results hold for all t.

Theorem 3 *Let $X = \{X(t), t \in \mathbf{R}^+\}$ be a real valued symmetric Lévy process with $\sigma^2(x)$ concave on $[0, \delta]$ and regularly varying at zero of order $\beta - 1$ where $1 < \beta \leq 2$, and let $\{L_t^x, (t,x) \in \mathbf{R}^+ \times \mathbf{R}\}$ be the local time of X.*

Let $\Phi(x)$ denote any function which is an inverse for $\sigma(x)$ near $x = 0$. (Thus, $\Phi(x)$ is regularly varying at zero of order $2/(\beta - 1)$). If $\{\pi(n)\}$ is any sequence of partitions of $[0, a]$ with $\lim_{n\to\infty} m(\pi(n)) = 0$ then

$$\lim_{n\to\infty} \sum_{x_i \in \pi(n)} \Phi(|L_t^{x_i} - L_t^{x_{i-1}}|) = c(\beta) \int_0^a |L_t^x|^{1/(\beta-1)}\, dx \tag{1.12}$$

in L^r uniformly in t on any bounded interval of R^+, for all $r > 0$, where $c(\beta)$ is given in (1.6).

In Section 2 we give the proof of Theorem 2, and in Section 3 we give the proof of Theorem 3. In Section 4 we provide a large class of examples of Lévy processes satisfying the assumptions of our Theorem 1. In particular, we show in Corollary 1 that if $h(x)$ is any function which is regularly varying and increasing as $x \to \infty$, and if $1 < \beta \le 2$, then we can find a Lévy process with $\sigma^2(x)$ concave such that $\sigma^2(x) \sim |x|^{\beta-1} h(\ln 1/x)$ as $x \to 0$.

2 Almost Sure Variation of Gaussian Processes

In this section we prove Theorem 2 which gives almost sure variation results for a wide class of Gaussian processes. The basic ideas are already contained in the proof of Theorem 1.2 of [4]. The main point of this section is to show how t hat proof can be adapted to the Orlicz space setting.

<u>**Proof of Theorem 2:**</u> Set $p = 2/(\beta - 1)$. It will suffice to prove the theorem for $\Phi(x)$ chosen to be a polynomially bounded Young's function regularly varying at zero of order $p \ge 2$. We first show that

$$\lim_{n\to\infty} E\left(\sum_{x_i \in \pi(n)} \Phi(G_{x_i} - G_{x_{i-1}})\right) = E(|Z|^p)a. \tag{2.1}$$

where Z is a standard Gaussian random variable. Let $Z_i = \frac{G_{x_i} - G_{x_{i-1}}}{\sigma(x_i - x_{i-1})} \stackrel{d}{=} N(0,1)$ and write

$$E\left(\sum_{x_i \in \pi(n)} \Phi(G_{x_i} - G_{x_{i-1}})\right) = E\left(\int_0^a F_n(\omega, x)\, dx\right) \tag{2.2}$$

where

$$\begin{aligned} F_n(\omega, x) &= \textstyle\sum_i \frac{\Phi(G_{x_i} - G_{x_{i-1}})}{\Delta x_i} 1_{(x_{i-1}, x_i]}(x) \\ &= \textstyle\sum_i \frac{\Phi(Z_i \sigma(\Delta x_i))}{\Phi(\sigma(\Delta x_i))} 1_{(x_{i-1}, x_i]}(x) \end{aligned} \tag{2.3}$$

and $\Delta x_i = x_i - x_{i-1}$. We begin by showing that

$$\|F_n\|_2^2 \le c \tag{2.4}$$

uniformly in $L^2(\Omega \times [0, a], P \times dx)$. This inequality, (2.4), will be used later on, in (2.24), and the method used to prove it will lead to (2.1).

To prove (2.4) we first note that

$$F_n^2 = \sum_i \left(\frac{\Phi(Z_i \sigma(\Delta x_i))}{\Phi(\sigma(\Delta x_i))}\right)^2 1_{(x_{i-1}, x_i]}(x). \tag{2.5}$$

If $|Z_i| \le 1$, then by monotonicity

$$\frac{\Phi(Z_i \sigma(\Delta x_i))}{\Phi(\sigma(\Delta x_i))} \le 1. \tag{2.6}$$

If $|Z_i| \geq 1$ is such that for suitable $\delta > 0$ we have $|Z_i|\sigma(\Delta x_i) \leq \delta$, then

$$\frac{\Phi(Z_i\sigma(\Delta x_i))}{\Phi(\sigma(\Delta x_i))} \leq c|Z_i|^{p+\epsilon} \tag{2.7}$$

by the regular variation of Φ. See e.g. [1], Theorem 1.5.6 (iii). Finally, if $|Z_i|\sigma(\Delta x_i) \geq \delta$, we can use the assumption that Φ is polynomially bounded to see that for some c and $k > p+1$

$$\begin{aligned} \frac{\Phi(Z_i\sigma(\Delta x_i))}{\Phi(\sigma(\Delta x_i))} &\leq c|Z_i|^k \frac{\sigma^k(\Delta x_i)}{\Phi(\sigma(\Delta x_i))} \\ &\leq c|Z_i|^k. \end{aligned} \tag{2.8}$$

Using (2.6)-(2.8) in (2.5) we see that

$$F_n^2 \leq c\sum_i \left(1+|Z_i|^k\right)^2 1_{(x_{i-1},x_i]}(x). \tag{2.9}$$

and (2.4) follows, since $Z_i \stackrel{d}{=} N(0,1)$.

We get (2.1) from the same reasoning: Note that we can write

$$E\left(\sum_{x_i\in\pi(n)} \Phi(G_{x_i} - G_{x_{i-1}})\right) = E\left(\int_0^a \tilde{F}_n(\omega,x)\,dx\right) \tag{2.10}$$

where now

$$\tilde{F}_n(\omega,x) = \sum_i \frac{\Phi(Z\sigma(\Delta x_i))}{\Phi(\sigma(\Delta x_i))} 1_{(x_{i-1},x_i]}(x). \tag{2.11}$$

i.e. we replace Z_i by Z. As before, in the proof of (2.4), we find that $\tilde{F}_n$ is uniformly bounded in L^2 , hence uniformly integrable. But by the regular variation of Φ we have

$$\lim_{n\to\infty} \tilde{F}_n(\omega,x) = |Z(\omega)|^p \qquad a.s.$$

Therefore we see that (2.1) follows from (2.10).

With these preliminaries out of the way, the proof of Theorem 2 will proceed analogously to the proof of Theorem 1.2 in [4]. Instead of l_p we will use the Orlicz space l_Φ. Let f, g be sequences of real numbers, $f = \{f_i\}, g = \{g_i\}, i = 1, 2, \ldots, \infty$. Recall the definition of the Luxembourg norm:

$$\|f\|_{(\Phi)} = \inf\{c : \sum_i \Phi\left(\frac{f_i}{c}\right) \leq 1\} \tag{2.12}$$

and note that

$$\|f\|_{(\Phi)} = \sup_{\{g:\|g\|_{\Phi^*}\} \leq 1} \sum_i f_i g_i \tag{2.13}$$

see [3], (14.10). In the last equation, Φ^* is the dual Young's function, and $||g||_{\Phi^*}$ is the 'standard' norm which satisfies

$$(2.14) \qquad ||g||_{\Phi^*} \leq 1 \Rightarrow \sum_i \Phi^*(g_i) \leq 1,$$

see e.g. [3], Lemma 9.2.

We use the notation $\Delta G(\pi)$ to denote the sequence $\{G_{x_i} - G_{x_{i-1}}, i = 1, 2, \ldots, k_\pi\}$, and for any sequence g we set $\langle g, \Delta G(\pi)\rangle = \sum_{i=1}^{k_\pi} g_i(G_{x_i} - G_{x_{i-1}})$. Let

$$M_n = \text{median}\left(||\Delta G(\pi(n))||_{(\Phi)}\right).$$

Then, using Borell's inequality as we did in lemma 2.1 of [4], we have

$$(2.15) \qquad P\left(|\, ||\Delta G(\pi(n))||_{(\Phi)} - M_n \,| > t\right) \leq 2\exp(-t^2/2\hat{\sigma}_n^2)$$

and the estimate

$$(2.16) \qquad |\, E\left(||\Delta G(\pi(n))||_{(\Phi)}\right) - M_n \,| \leq \hat{\sigma}_n\sqrt{2\pi}$$

where

$$\hat{\sigma}_n^2 \stackrel{\text{def}}{=} \sup_{\{g:||g||_{\Phi^*} \leq 1\}} E\left\{\langle g, \Delta G(\pi(n))\rangle^2\right\}.$$

Using (2.14) we see that

$$\begin{aligned} \hat{\sigma}_n^2 &\leq \sup_{\{g:\sum_i \Phi^*(g_i)\leq 1\}} E\left\{\langle g, \Delta G(\pi(n))\rangle^2\right\} \\ &\leq \sup_{\{g:||g||_2\leq \tilde{c}\}} E\left\{\langle g, \Delta G(\pi(n))\rangle^2\right\} \\ (2.17) \qquad &\leq c\sup_i \sigma^2(x_i - x_{i-1}) \end{aligned}$$

for some $c < \infty$, as in the proof of lemma 2.2 of [4]. Here we have used the fact that for some $\tilde{c} < \infty$

$$\Phi^*(x) \leq 1 \Rightarrow x^2 \leq \tilde{c}^2\Phi^*(x).$$

This fact is most easily derived from the following chain of implications for x near zero: By the concavity of σ^2, we have $\sigma^2(x) \geq cx$ and therefore $\sigma(x) \geq cx^{1/2}$, so that $\Phi(x) \leq cx^2$, which implies that $\Phi^*(x) \geq cx^2$ b y (2.9) of [3].

Note that

$$||\Delta G(\pi(n))||_{(\Phi)} \leq 1 + \sum_i \Phi(G_{x_i} - G_{x_{i-1}}),$$

see e.g. [3], (9.12) and (9.20). Using this, (2.1), (2.16) and (2.17) show that $\{M_n\}, n = 1, \ldots, \infty$ is bounded. Hence some subsequence M_{n_j} has a limit point M_*. We now show that

$$(2.18) \qquad M_*^p = E(|Z|^p)a$$

and that along the subsequence $\{n_j\}$

$$\lim_{j\to\infty} \sum_{x_i\in\pi(n_j)} \Phi(G_{x_i}-G_{x_{i-1}}) = M_*^p \qquad a.s. \tag{2.19}$$

This will complete the proof of our theorem just as in the proof of Theorem 1.2 of [4], since the uniqueness of the limit point M_* demonstrated in (2.18) will show that we actually have convergence of the full sequence in (2.19).

To prove (2.18) and (2.19) we begin by arguing exactly as in [4] to find that

$$R_{n_j}(\omega) \stackrel{\text{def}}{=} \|\Delta G(\pi(n_j))\|_{(\Phi)} \to M_* \qquad a.s. \quad \text{as } j\to\infty \tag{2.20}$$

By definition we have

$$\sum_{x_i\in\pi(n_j)} \Phi\left(\frac{G_{x_i}(\omega)-G_{x_{i-1}}(\omega)}{R_{n_j}(\omega)}\right) = 1 \tag{2.21}$$

see e.g. [3], chapter II, section 7. Now note that by the regular variation of Φ, for fixed ω in a set of measure 1 and for any $\epsilon > 0$, we can find N so large that for all $n_j \geq N$ we have

$$\begin{aligned} &\frac{(1-\epsilon)}{M_*^p} \sum_{x_i\in\pi(n_j)} \Phi(G_{x_i}-G_{x_{i-1}}) \\ &\leq \sum_{x_i\in\pi(n_j)} \Phi\left(\frac{G_{x_i}-G_{x_{i-1}}}{R_{n_j}}\right) \\ &\leq \frac{(1+\epsilon)}{M_*^p} \sum_{x_i\in\pi(n_j)} \Phi(G_{x_i}-G_{x_{i-1}}). \end{aligned} \tag{2.22}$$

Therefore by (2.21) and (2.22) we get (2.19). To obtain (2.18) we observe that in the notation of (2.3)

$$\sum_{x_i\in\pi(n_j)} \Phi(G_{x_i}-G_{x_{i-1}}) = \int_0^a F_{n_j}(\omega,x)\,dx. \tag{2.23}$$

Hence, using (2.4) and Cauchy-Schwarz inequality we see that

$$\begin{aligned} E\left[\left(\sum_{x_i\in\pi(n_j)} \Phi(G_{x_i}-G_{x_{i-1}})\right)^2\right] &= E\left[\left(\int_0^a F_{n_j}(x)\,dx\right)^2\right] \\ &\leq aE\left(\int_0^a F_{n_j}^2(x)\,dx\right) \\ &\leq ca \end{aligned} \tag{2.24}$$

uniformly in n_j. Therefore, (2.23) is uniformly integrable, and (2.19) and (2.1) now imply that

$$\begin{aligned} M_*^p &= \lim_{j\to\infty} E\left(\textstyle\sum_{x_i\in\pi(n_j)} \Phi(G_{x_i} - G_{x_{i-1}})\right) \\ &= E(|Z|^p)a \end{aligned} \tag{2.25}$$

by (2.1). This proves (2.18), and completes the proof of (1.10) of Theorem 2. (1.11) follows from (1.10) as in the proof of Theorem 1.2 of [4].

3 L^r Convergence

Theorem 3 will follow from Theorem 1 as in the proof of Theorem 1.1 of [4] once we establish the next lemma, which is the analogue of Lemma 3.4 of [4].

For a fixed partition π we introduce the notation

$$\|L_t\|_{\pi,\Phi} = \sum_{x_i\in\pi} \Phi(\Delta L_t^{x_i})$$

where

$$\Delta L_t^{x_i} = L_t^{x_i} - L_t^{x_{i-1}},$$

and we let

$$g(t) = \int_0^t p_s(0)\, ds$$

denote the partial Greens' function.

Lemma 1 *Let $X = \{X(t), t \in \mathbf{R}^+\}$ be a real valued symmetric Lévy process with $\sigma^2(x)$ concave on $[0,\delta]$ and regularly varying at zero of order $\beta-1$ where $1 < \beta \le 2$, and let $\{L_t^x, (t,x) \in \mathbf{R}^+ \times \mathbf{R}\}$ be the local time of X.*

Let $\Phi(x)$ denote any function which is an inverse for $\sigma(x)$ near $x = 0$. (Thus, $\Phi(x)$ is regularly varying at zero of order $2/(\beta-1)$). Then we can find an $\epsilon > 0$ such that for all partitions π of $[0,a]$ with $m(\pi) \le \epsilon$, and all $s,t,T \in R^+$, with $s \le t \le T < \infty$, and integers $m \ge 1$

$$\| \, \|L_t\|_{\pi,\Phi} - \|L_s\|_{\pi,\Phi} \|_m \le C(\beta,m,T) g^{1/2}(t-s)a \tag{3.1}$$

and in particular

$$\| \, \|L_t\|_{\pi,\Phi} \|_m \le C(\beta,m,T) g^{1/2}(t)a. \tag{3.2}$$

Proof: As before, we can take Φ to be convex, so that the monotonicity of Φ' together with the Mean Value Theorem show that

$$|\Phi(u) - \Phi(v)| \le (\Phi'(u) + \Phi'(v))\, |u - v|.$$

Hence

$$(3.3)\quad \begin{aligned} &\| \, \|L_t\|_{\pi,\Phi} - \|L_s\|_{\pi,\Phi}\|_m \\ &\leq \sum_{x_i \in \pi} \|\Phi(\Delta L_t^{x_i}) - \Phi(\Delta L_s^{x_i})\|_m \\ &\leq \sum_{x_i \in \pi} (\|\Phi'(\Delta L_t^{x_i})\|_{2m} + \|\Phi'(\Delta L_s^{x_i})\|_{2m})\|\Delta L_t^{x_i} - \Delta L_s^{x_i}\|_{2m} \end{aligned}$$

With the notation

$$Z_i = \frac{\Delta L_t^{x_i}}{\sigma(\Delta x_i)}$$

and using (2.6)-(2.8), but with Φ replaced by Φ' we see that for some fixed c and any $k > p = 2/(\beta - 1)$

$$(3.4)\quad \begin{aligned} \|\Phi'(\Delta L_t^{x_i})\|_{2m} &= \|\tfrac{\Phi'(Z_i\sigma(\Delta x_i))}{\Phi'(\sigma(\Delta x_i))}\|_{2m}\Phi'(\sigma(\Delta x_i)) \\ &\leq c\|1 + Z_i^k\|_{2m}\Phi'(\sigma(\Delta x_i)). \end{aligned}$$

By (3.17) of [4] we see that for any $u \in R$ and any integer $j \geq 1$

$$(3.5)\quad E^u(\Delta L_t^{x_i})^{2j} \leq (2j)!2^j(g^{1/2}(t)\sigma(\Delta x_i))^{2j}$$

so that

$$(3.6)\quad \|1 + Z_i^k\|_{2m} \leq 1 + cg^{k/2}(t).$$

Similarly, using (3.5) together with the Markov property as in the proof of lemma 3.4 of [4] we see that

$$(3.7)\quad \|\Delta L_t^{x_i} - \Delta L_s^{x_i}\|_{2m} \leq cg^{1/2}(t-s)\sigma(\Delta x_i).$$

Using once again the convexity of Φ, the Monotone Density Theorem, (theorem 1.7.2b of [1]) shows that

$$(3.8)\quad x\Phi'(x) \sim p\Phi(x) \qquad \text{as} \quad x \to 0.$$

Putting these estimates together and remembering that Φ is an inverse for σ near zero finishes the proof of Lemma 1.

4 Examples

We will now give examples of Lévy processses which satisfy the assumptions of Theorem 1.

Lemma 2 *Let μ be a finite positive measure on* $(1, 2]$ *and let*

$$\psi(\lambda) = \int_1^2 \lambda^s \, d\mu(s).$$

Then ψ is a Lévy exponent, and

$$\sigma^2(x) = \int_0^\infty \frac{(1-\cos\lambda x)}{\psi(\lambda)}\, d\lambda$$

is concave on $[0,\infty)$.

Proof: Note that

$$\begin{aligned}(4.1)\quad & 2\sigma^2(x) - \sigma^2(x-h) - \sigma^2(x+h) \\ &= \int_0^\infty (1-\cos v) \\ &\quad \left[\frac{2}{x\psi(v/x)} - \frac{1}{(x+h)\psi(v/x+h)} - \frac{1}{(x-h)\psi(v/x-h)}\right] dv\end{aligned}$$

for all $|h| \le |x|$. Therefore, to show that σ^2 is concave on $[0,\infty)$ it suffices to show that the term in the bracket is positive, i.e. that $1/x\psi(v/x)$ is concave in x for all $x > 0, v > 0$. This is clearly equivalent to showing that $g(x) = 1/x\psi(1/x)$ is concave for $x > 0$.

By definition

$$g(x) = \frac{1}{\int_1^2 x^{1-s}\, d\mu(s)}$$

so that

$$g'(x) = \frac{\int_1^2 (s-1)x^{-s}\, d\mu(s)}{\left(\int_1^2 x^{1-s}\, d\mu(s)\right)^2}$$

and

$$g''(x) = \frac{2\left(\int_1^2 (s-1)x^{-s}\, d\mu(s)\right)^2}{\left(\int_1^2 x^{1-s}\, d\mu(s)\right)^3} - \frac{\int_1^2 s(s-1)x^{-s-1}\, d\mu(s)}{\left(\int_1^2 x^{1-s}\, d\mu(s)\right)^2}.$$

Thus, $g'' \le 0$ if

$$2\left(\int_1^2 (s-1)x^{-s}\, d\mu(s)\right)^2 \le \int_1^2 s(s-1)x^{-s-1}\, d\mu(s) \int_1^2 x^{1-s}\, d\mu(s)$$

or equivalently

$$2\left(\int_1^2 (s-1)x^{-s}\, d\mu(s)\right)^2 \le \int_1^2 s(s-1)x^{-s}\, d\mu(s) \int_1^2 x^{-s}\, d\mu(s).$$

This last inequality follows from the Schwartz inequality applied to

$$\begin{aligned}&2\left(\int_1^2 (s-1)x^{-s}\, d\mu(s)\right)^2 \\ &= 2\left(\int_1^2 \left[x^{-s}(s-1)/s\right]^{1/2} \left[x^{-s}(s-1)s\right]^{1/2} d\mu(s)\right)^2\end{aligned}$$

since $2(s-1)/s \leq 1$.

That $\psi(\lambda)$ is a Lévy exponent follows immediately from the fact that λ^s is a Lévy exponent.□

Lemma 3 *Let $1 < \beta \leq 2$, and let $\rho(s)$ be a bounded, continuous, increasing function such that the measure $d\rho$ is supported on the interval $[0, \beta - 1)$. Then, we can find a Lévy process with exponent ψ such that*

$$\psi(\lambda) = \lambda^\beta \hat{\rho}(\ln \lambda)$$

for all $\lambda > 0$, and such that $\sigma^2(x)$ is concave on $[0, \infty)$. Here

$$\hat{\rho}(\lambda) = \int_0^\infty e^{-\lambda s}\, d\rho(s).$$

Proof: Set

$$\mu(s) = |\rho| - \rho(\beta - s)$$

where $|\rho|$ denotes the mass of $d\rho$. $\mu(s)$ is a continuous, increasing function such that the measure $d\mu$ is supported on the interval $(1, \beta]$. By Lemma 2, $\psi(\lambda) = \int_1^2 \lambda^s\, d\mu(s)$ is a Lévy exponent and the function σ^2 associated with ψ is concave on $[0, \infty)$.

Our present lemma then follows from the fact that

$$\text{(4.2)} \qquad \begin{aligned} \psi(\lambda) = \int_1^2 \lambda^s\, d\mu(s) &= -\int_1^\beta \lambda^s\, d\rho(\beta - s) \\ &= \lambda^\beta \int_0^{\beta-1} \lambda^{-s}\, d\rho(s) \\ &= \lambda^\beta \int_0^{\beta-1} e^{-(\ln \lambda)s}\, d\rho(s) \\ &= \lambda^\beta \hat{\rho}(\ln \lambda). \qquad \square \end{aligned}$$

Combining Lemma 3, (1.4), and Theorem 1.7.1' of [1], we obtain a large class of Lévy processes which satisfy the hypotheses of our theorems. In particular, if $\hat{\rho}(\ln \lambda)$ is slowly varying at infinity and $1 < \beta \leq 2$ then we can fin d a Lévy process with exponent ψ given by (3.2) and with concave $\sigma^2(x) \sim c_\beta |x|^{\beta-1} \frac{1}{\hat{\rho}(\ln 1/x)}$. By the cited Theorem in [1] we see that we can find a $\hat{\rho}(\lambda)$ asymptotic at infinity to any regular ly varying function of index less than zero, or to any decreasing slowly varying function. Taking $h(x) = c_\beta \frac{1}{\hat{\rho}(\ln 1/x)}$ leads to the following Corollary mentioned at the end of section 1.

Corollary 1. *Let $h(x)$ be any function which is regularly varying and increasing as $x \to \infty$, and let $1 < \beta \leq 2$. Then we can find a Lévy process with $\sigma^2(x)$ concave such that*

$$\sigma^2(x) \sim |x|^{\beta-1} h(\ln 1/x) \qquad \textit{as } x \to 0.$$

References

[1] N. Bingham, C. Goldie, and J. Teugals, *Regular Variation*, Cambridge University Press, Cambridge, 1987.

[2] T. Kawada and N. Kono, *On the variation of Gaussian processes*, Proceedings of the Second Japan-USSR Symposium on Probability Theory (Berlin), Lecture Notes Math, vol. 330, Springer-Verlag, Berlin, 1973, pp. 175–192.

[3] M. Krasnoselskii and Ya. Rutickii, *Convex functions and Orlicz spaces*, P. Noordhoff, Gronigen, 1961.

[4] M. Marcus and J. Rosen, *p-variation of the local times of symmetric stable processes and of Gaussian processes with stationary increments*, Preprint. To appear in the Annals of Probability, 1992.

[5] ———, *Sample path properties of the local times of strongly symmetric Markov processes via Gaussian processes*, Special Invited Paper. To appear in the Annals of Probability, 1992.

[6] E. J. G. Pitman, *On the behavior of the characteristic function of a probability distribution in the neighbourhood of the origin*, J. Australian Math. Soc. Series A **8** (1968), 422–443.

Michael B. Marcus
Department of Mathematics
The City College of CUNY
New York, NY 10031

Jay Rosen
Department of Mathematics
College of Staten Island, CUNY
Staten Island, NY 10301

CRITICAL RANDOM WALK IN RANDOM ENVIRONMENT ON TREES OF EXPONENTIAL GROWTH

ROBIN PEMANTLE [1] [2]

ABSTRACT:

This paper studies the behavior of RWRE on trees in the critical case left open in previous work. For trees of exponential growth, a random perturbation of the transition probabilities can change a transient random walk into a recurrent one. This is the opposite of what occurs on trees of sub-exponential growth.

1 Introduction

This paper is concerned with the problem of determining whether a random walk in a random environment (RWRE) on an infinite, exponentially growing tree is transient or recurrent. The problem was first studied in [8] as a way of analyzing another process called Reinforced Random Walk, and then in [7] where a more complete solution was obtained. It was shown there that the RWRE is transient when the size of the tree, as measured by the log of the *branching number*, is greater than the *backward push* of the random environment, and recurrent when the log of the branching number is smaller than the backward push. The case of equality was left open. For trees of sub-exponential growth, this critical case was almost completely settled in [9]. The present paper is a companion to [9] in that it attempts to settle the critical case for exponentially growing trees. The results here are less definitive than in the sub-exponential case, in that the sufficient

[1] Research supported in part by a National Science Foundation postdoctoral fellowship and be NSF grant # DMS9103738

[2] Department of Mathematics, University of Wisconsin-Madison, Van Vleck Hall, 480 Lincoln Drive, Madison, WI 53706

conditions for transience and for recurrence are in terms of capacity and growth respectively; these conditions are not quite complementary, leaving open a critical-within-critical case. A technical assumption on the random environment is also required; examples show that this assumption is often satisfied. On the positive side, it is proved here that a phase boundary occurs in an unusual place, namely when the growth rate of the tree, $\exp(\beta n + o(n))$, has the $o(n)$ term equal to a constant times $n^{1/3}$. Here follows a precise description of the problem.

Let Γ be any infinite, locally finite tree with no leaves (vertices of degree one). Designate a vertex ρ of Γ as its root. For any vertex $\sigma \neq \rho$, denote by σ' the unique neighbor of σ closer to ρ (σ' is also called the parent of σ). An environment for random walk on a fixed tree, Γ, is a choice of transition probabilities $q(\sigma, \tau)$ on the vertices of Γ with $q(\sigma, \tau) > 0$ if and only if σ and τ are neighbors. When these transition probabilities are taken as random variables, the resulting mixture of Markov chains is called Random Walk in Random Environment (RWRE). Following [7] and the references therein, random environments studied in this paper satisfy the homogeneity condition

$$\text{The variables } X(\sigma) = \ln\left(\frac{q(\sigma', \sigma)}{q(\sigma', \sigma'')}\right) \text{ are i.i.d. for } |\sigma| \geq 2, \tag{1}$$

where $|\sigma|$ denotes the distance from σ to ρ. Here, and throughout, let X denote a random variable with this common distribution.

Before stating the main result, a few definitions and notations are required. Write $\sigma \leq \tau$ if σ is on the path connecting ρ and τ; in this paper, the term "path" always refers to a path without self-intersection. Write $\sigma \wedge \tau$ for the greatest lower bound of σ and τ; pictorially, this is where the paths from ρ to σ and τ diverge. Let $\partial\Gamma$, called the ***boundary*** of Γ, denote the set of infinite paths beginning at ρ. Let Γ_n denote the set $\{\sigma : |\sigma| = n\}$ of vertices at level n of Γ. Define the ***backward push*** of the random environment, denoted β or $\beta(X)$, by

$$\beta(X) = -\ln \min_{0 \leq \lambda \leq 1} \mathbf{E}e^{\lambda X}.$$

The size of an infinite tree is best discussed in terms of capacity.

Definition 1 *Let $\phi : \mathbf{Z}^+ \to \mathbf{R}^+$ be a nonincreasing function. Define the*

ϕ-energy of a probability measure μ on the boundary of Γ to be

$$I_\phi(\mu) = \int_{\partial\Gamma}\int_{\partial\Gamma} \phi(|\xi \wedge \eta|)^{-1}\, d\mu(\xi)\, d\mu(\eta).$$

Define the capacity of Γ in gauge ϕ by

$$Cap_\phi(\Gamma) = \left[\inf_\mu I_\phi(\mu)\right],$$

where the infimum is over all probability measures on $\partial\Gamma$., and $Cap_\phi(\Gamma) \neq 0$ if and only if there is some measure of finite energy.

Say that Γ is *spherically symmetric* if there is a *growth function* $f : \mathbf{Z}^+ \to \mathbf{Z}^+$ such that every vertex $\sigma \neq \rho$ has $1+f(|\sigma|)$ neighbors; in other words, the degree of a vertex depends only on its distance from the root. A spherically symmetric tree Γ has positive capacity in gauge ϕ if and only if

$$\sum \phi(n)|\Gamma_n|^{-1} < \infty.$$

Thus positive capacity in gauge $\phi(n) = e^{-kn}$ implies liminf exponential growth rate of at least k. In particular, the supremum of those k for which Γ has positive capacity in gauge $\phi(n) = e^{-kn}$ is the Hausdorff dimension, $dim(\Gamma)$; in the terminology of [5] and [7], $dim(\Gamma)$ is the log of the branching number.

The main result of [7] is that RWRE on Γ is a.s. transient if $dim(\Gamma) > \beta(X)$, and a.s. recurrent if $dim(\Gamma) < \beta(X)$. The use of gauges more general than e^{-kn} allows for finer distinctions of size to be made within the class of trees of the same dimension. In particular, when $\beta = dim(\Gamma) = 0$, it is shown in [9] that positive capacity in gauge $n^{1/2}$ is sufficient and almost necessary for transience of RWRE. Since the case $\beta = 0$ is in some sense a mean-zero perturbation of the deterministic environment of a simple random walk ($X \equiv 0$), and simple random walk is transient if and only if Γ has positive capacity in gauge n^{-1}, this shows that the perturbation makes the walk more transient. By contrast, the main result of this paper is as follows.

Definition 2 *Say that a real random variable X is* top-heavy *if the infimum of $\mathbf{E}e^{\lambda X}$ over $\lambda \in [0,1]$ is achieved at some $\lambda_0 \in (0,1)$ and $\mathbf{E}e^{\gamma X} < \infty$ for some $\gamma > \lambda_0$.*

Theorem 1 *Consider RWRE on a tree* Γ *with* $\beta(X) = dim(\Gamma) > 0$. *If* $X = -\beta$ *with probability one, then RWRE is transient if and only if* Γ *has positive capacity in gauge* $\phi(n) = e^{-n\beta}$. *On the other hand if* X *is nondeterministic, top-heavy, and is either a lattice distribution or has an absolutely continuous component with density bounded above and bounded away from zero in a neighborhood of zero, then*

(*i*) *there exists* $c_1(X)$ *for which the growth bound*

$$|\Gamma_n| \leq e^{\beta n + c_1 n^{1/3}} \text{ for all } n$$

implies that RWRE is recurrent;

(*ii*) *there exists* $c_2(X)$ *such that if* Γ *has positive capacity in gauge* $\phi(n) = e^{-n\beta - c_2 n^{1/3}}$ *then RWRE is transient.*

Remark: The requirement that X be top-heavy is enigmatic, but not overly restrictive. For example, it is satisfied by normal random variables with mean $-c$ and variance V whenever $c < 2V$. In the case where X takes only the values ± 1, with $\mathbf{P}(X = 1) = p < 1/2$, it is top-heavy if and only if $(1-p)/p < e^2$.

The remainder of this section outlines the the proof of this theorem and serves as a guide to the remaining sections. Theorem 1 is proved in the following three steps. First, in Section 2, a correspondence connection between random walks and electrical networks [2] reduces the problem to one of determining whether a random electrical network is transient or recurrent almost surely. After this reduction, the technical condition of top-heaviness comes in: top-heaviness implies that finite resistance, when achieved, will be due to a single random infinite path of finite resistance; searching for a single path with this property is easier than searching for some large collection of paths with a weaker property. Next, large deviation estimates are needed for the probability of an unusually small resistance along a fixed path of length n (Lemma 6 and Corollary 7). These are applied via a simple first-moment calculation to obtain Lemma 8, which is an upper bound tending to zero on the probability that any of the $|\Gamma_n|$ chains of resistances of length n stays small. Incidentally, this is where the "extra" factor of $e^{cn^{1/3}}$ comes in. Estimates with this same factor have been obtained by Kesten with much greater accuracy for branching Brownian motion [4]. Ours are a discrete analogue of Kesten's in the sense that continuous-time branching has been replaced by Γ-indexed branching; this analogy is explained more fully in [1] and [9]. Part (*i*) of Theorem 1 follows from this upper bound by computing the expected truncated conductance. The last step, which is needed only for the proof of part (*ii*), is a second-moment technique (Lemma 3) developed in [6] and [9] for proving the almost sure existence of an infinite path of finite resistance based on the two-dimensional marginals for finite paths (i.e. the probabilities for two paths of length n both to have large conductance if the paths share the first k resistors).

2 Reductions

Begin with the reduction of the recurrence/transience problem to an electrical problem. As is well known, transience of a reversible Markov chain is equivalent to finite resistance of the associated resistor network on the same graph, where the transition probabilities from any vertex are proportional to the conductances (reciprocal resistances) of the edges incident to that vertex; see for example [2]. For a random environment satisfying (1), the resistances in the associated random electrical network are easily seen to be given by

$$\text{Resistance along } \overline{\sigma'\sigma} = e^{-S(\sigma)}$$

where

$$S(\sigma) = \prod_{\rho<\tau\leq\sigma} X(\sigma).$$

Here, the values of $X(\sigma)$ for $|\sigma| \leq 1$ are assigned to make this relation hold for all $\sigma \neq \rho$, while the values for $|\sigma| \geq 2$ are i.i.d. by (1). Since finiteness of the total resistance is not affected by changing finitely many resistances, we alter the $X(\sigma)$ for $|\sigma| \leq 1$ so that the entire collection is i.i.d.

A sufficient condition for transience is the existence of an infinite path $\rho, \sigma_1, \sigma_2, \ldots$ along which $\sum e^{-S(\sigma)} < \infty$. Conversely, let

$$U(\sigma) = \min_{\rho<\tau\leq\sigma} e^{S(\sigma)}.$$

A useful sufficient condition for recurrence is given by the following lemma.

Lemma 2 *Let Γ be any tree with conductances $C(\sigma)$, and let Π be any cutset, i.e. any minimal set among those intersecting every infinite path from ρ. Then the conductance from ρ to Π is at most*

$$\sum_{\sigma \in \Pi} U(\sigma).$$

Consequently, if the conductances are random with $\sum_{|\sigma|=n} U(\sigma) \to 0$ in probability, then the random walk is recurrent with probability one.

Proof: For each $\sigma \in \Pi$, let $\gamma(\sigma)$ be the sequence of conductances on the path from ρ to σ, and let Γ' be a tree consisting of disjoint paths for each $\sigma \in \Pi$, each path having conductances $\gamma(\sigma)$. Γ is a contraction of Γ', so by Rayleigh's monotonicity law, the conductance to Π in Γ is less than or equal to the conductance of Γ', which is the sum over $\sigma \in \Pi$ of conductances bounded above by $U(\sigma)$. Putting $\Pi = \Gamma_n$ shows that the conductance from ρ to infinity is bounded above by $\liminf_n \sum_{|\sigma|=n} U(\sigma)$, proving the lemma. □

In the next section estimates will be given on $\mathbf{P}(\sigma \in W)$ and $\mathbf{P}(\sigma, \tau \in W)$, where for fixed constants c and L, W is the set of vertices σ such that for every $\tau \leq \sigma$ with $|\tau| > L$,

$$1/10 \leq S(\tau)/c|\tau|^{1/3} \leq 1.$$

These estimates are then plugged into the following result of Lyons (see [9, Theorem 4.1]).

Lemma 3 *Let Γ be an infinite, locally finite tree without leaves, let $X(\sigma)$ be i.i.d. random variables indexed by the vertices of Γ and let B_n be a subset of $\mathbf{R}^n$ for each n. Let W be the set of vertices $\sigma \in \Gamma$ such that for every $\tau \leq \sigma$, the sequence $(X(\rho), \ldots, X(\tau))$ along the path from ρ to τ is in the set $B_{|\tau|}$. Suppose there is a positive, nonincreasing function $g : \mathbf{Z}^+ \to \mathbf{R}$ such that for any two vertices $\sigma, \tau \in \Gamma_n$ with $|\sigma \wedge \tau| = k$,*

$$\mathbf{P}(\sigma, \tau \in W) \leq \frac{\mathbf{P}(\sigma \in W)^2}{g(k)}. \tag{2}$$

Then the probability of W containing an infinite path is at least $Cap_g(\Gamma)$. □

Remark: Of course any infinite path in W has $\sum e^{-S(\sigma)} < \infty$, implying transience. The reason that one looks for a path along which $S(\tau)$ is bounded above as well as below is so as to be able to apply this lemma, which is really a jazzed up second moment bound. It is important to find a random subset of Γ_n whose cardinality has a second moment not too much larger than the square of its first moment; then the LHS of (2) will not be too large.

3 Large deviation estimates

The first series of estimates concern the probabilities $\mathbf{P}(\sigma \in W)$ of the previous lemma. The routine proofs of the first two propositions in the series are omitted.

Proposition 4 *Let $\mathbf{P}_x$ be the law of a standard one-dimensional Brownian motion started at x. Let $-1 < c_1 < c_2 < 1$ and $-1 < c_3 < c_4 < 1$ be real constants. Then there exist positive constants K_1 and K_2 such that for any $L > 0$, the following two inequalities hold.*

$$\sup_x \mathbf{P}_x(|B_t| \leq 1 \textit{ for all } t \leq L) \quad \leq \quad K_1 e^{-\frac{\pi^2}{8}L} \quad ;$$

$$\inf_{c_1 \leq x \leq c_2} \mathbf{P}_x(|B_t| \leq 1 \textit{ for all } t \leq L \textit{ and } c_3 \leq |B_L| \leq c_4) \quad \geq \quad K_2 e^{-\frac{\pi^2}{8}L}.$$

Proposition 5 *Let $\{X_n\}$ be a sequence of i.i.d. random variables with mean zero and variance $V < \infty$. Let $S_n = \sum_{i=1}^n X_i$ and for $0 < t < 1$, interpolate polygonally by letting $S_{n+t} = S_n + tX_{n+1}$. Then*

$$\limsup_{L\to\infty} \limsup_{n\to\infty} \sup_{|x|\leq 1} L^{-1} \ln \Big(\mathbf{P}(|(Vn)^{-1/2} S_{nt} + x| \leq 1 \textit{ for all } t \leq L)\Big)$$

$$\leq -\frac{\pi^2}{8}. \tag{3}$$

If $-1 < c_1 < c_2 < 1$ and $-1 < c_3 < c_4 < 1$ are real constants, then

$$\liminf_{L\to\infty} \liminf_{n\to\infty} \inf_{c_1\leq x\leq c_2} L^{-1} \ln \mathbf{P}[A(L,n,x)]$$

$$\geq \; -\frac{\pi^2}{8}, \tag{4}$$

where $A(L,n,x)$ is the event

$$\left\{|(Vn)^{-1/2}S_{nt}+x|\leq 1 \text{ for all } t\leq L, \text{ and } c_3\leq S_{nL}+x\leq c_4\right\}.$$

Lemma 6 *Suppose $f,g:\mathbf{Z}^+\to\mathbf{R}$ satisfy $f>g$ and $\lim_{t\to\infty}f(t)-g(t)=\infty$, and assume the following flatness hypothesis:*

$$\sup_L \lim_{t\to\infty}\frac{\sup_{0\leq s\leq L(f(t)-g(t))^2}\max(|f(t+s)-f(t)|,|g(t+s)-g(t)|)}{f(t)-g(t)}=0.$$

Let $S_n=\sum_{i=1}^n X_i$ be a random walk with $\mathbf{E}X_1=0$ and $\mathbf{E}X_1^2=V<\infty$ and suppose that for each n, $\mathbf{P}(g(k)<S_k<f(k) \text{ for all } k\leq n)>0$. Then

$$\lim_{n\to\infty}\left(\sum_{k=1}^n(f(k)-g(k))^{-2}\right)^{-1}\ln\left(\mathbf{P}(g(k)<S_k<f(k) \text{ for all } k\leq n)\right)$$

$$=-\frac{\pi^2}{8}V. \tag{5}$$

Corollary 7 *Let Y_n be i.i.d. mean zero random variables with partial sums $T_n=\sum_{i=1}^n Y_i$ and suppose that $\mathbf{E}Y_1^2=V<\infty$. Then for any real $c_1<c_2$,*

$$\lim_{n\to\infty}\frac{\ln\mathbf{P}\left(c_1k^{1/3}\leq T_k\leq c_2k^{1/3} \text{ for all } k\leq n\right)}{n^{1/3}}=-\frac{\pi^2}{8}\frac{3V}{(c_2-c_1)^2}$$

unless the probability is eventually zero.

Proofs: The first proposition is a standard eigenvalue estimate – see for example [3, p. 259]. The second follows from the first and the invariance principle by piecing together segments of path of length L; the two-page proof is entirely routine and is omitted.

To establish the lemma, prove first that the limsup is at most $-\frac{\pi^2}{8}V$. Extend f, g and S to $\mathbf{R}^+$ by linear interpolation with $f(0)=g(0)=S_0=0$; for any integer n, the event $\{g(k)\leq S_k\leq f(k):k=1,\ldots,n\}$ is the same when k is replaced by a real parameter t running from 0 to n. Fix $L>0$. For any positive integer m_0 inductively define $m_{k+1}=\lceil m_k+(1/4)V^{-1}L(f(m_k)-g(m_k))^2\rceil$. Let $M(n)$ denote $\max\{k:m_k<n\}$.

Claim: $\lim_{n\to\infty}M(n)^{-1}\sum_{k=1}^n 4VL^{-1}(f(k)-g(k))^{-2}=1$.

To prove this, first choose $\epsilon\in(0,1/8)$ and j large enough so that $t\geq m_j$ implies

$$\sup_s\frac{\max(|f(t+s)-f(t)|,|g(t+s)-g(t)|)}{f(t)-g(t)}<\epsilon, \tag{6}$$

where the supremum is over s for which $0 \leq s \leq (1/4)V^{-1}L(f(t)-g(t))^2$. Summing the identity $\sum_{i=m_k}^{m_{k+1}}(m_{k+1}-m_k)^{-1} = 1$ over $j \leq k \leq M(n)+1$ yields

$$M(n) \leq j+1+\sum_{i=m_0}^{n} 4VL^{-1}(f(m_{M(i)})-g(m_{M(i)}))^{-2}$$

$$\leq j+1+\sum_{i=m_0}^{n} 4VL^{-1}(f(i)-g(i))^{-2}(1-2\epsilon)^{-2}.$$

Similarly,

$$M(n) \geq j+\sum_{i=m_0}^{n} 4VL^{-1}(f(i)-g(i))^{-2}(1+2\epsilon)^{-2}.$$

Letting $n \to \infty$ proves that the limsup and liminf of

$$M(n)^{-1}\sum_{k=1}^{n} 4VL^{-1}(f(k)-g(k))^{-2}$$

are between $(1+2\epsilon)^{-1}$ and $(1-2\epsilon)^{-1}$. Letting $\epsilon \to 0$ proves the claim.

Continuing the proof of the lemma, pick $\epsilon > 0$ and N large enough so that for our (still fixed) value of L and $n \geq N$, the sup in (3) is at most $-\frac{\pi^2}{8}+\epsilon$. Pick m_0 large enough so that $t \geq m_0$ implies firstly (6) and secondly $f(t)-g(t) > (4VN)^{1/2}$. Let $y_k = (f(m_k)+g(m_k))/2$ and $d_k = (f(k)-g(k))/2$, so $m_{k+1}-m_k = \lceil LV^{-1}d_k^2 \rceil$. Let R_k be the rectangle

$$\{(x,y) : m_k \leq x \leq m_{k+1} \text{ and } y_k-(1+2\epsilon)d_k \leq y \leq y_k+(1+2\epsilon)d_k\}.$$

Then the union of the R_k covers the graphs of f and g after m_0, i.e. the region $\{(x,y) : x \geq m_0, g(x) \leq y \leq f(x)\}$. Write S_i' for a random walk independent of S_i and identically distributed. Using the Markov property of S_i, the stationarity of its increments, and then rescaling each R_k to have height one, gives

$$\mathbf{P}(f(k) \leq S_k \leq g(k) : 1 \leq k \leq n)$$

$$\leq \mathbf{P}((i,S_i) \in \bigcup R_k : m_0 \leq i \leq n)$$

$$\leq \prod_{k=0}^{M(n)-1} \mathbf{P}\left((i,S_i) \in R_{k-1} : m_k \leq i \leq m_{k+1} \,|\, S_{m_k}\right)$$

$$= \prod_{k=1}^{M(n)-1} \mathbf{P}(|S_{m_k} + S'_i - y_k| \leq (1+2\epsilon)d_k : i = 1, \ldots, m_{k+1} - m_k)$$

$$= \prod_{k=1}^{M(n)-1} \mathbf{P}\Big(|(1+2\epsilon)^{-1} d_k^{-1}[(S_{m_k} - y_k) + S'_{V^{-1}(1+2\epsilon)^2 d_k^2 t}]| \leq 1$$
$$\text{for all } 0 \leq t \leq L(1+2\epsilon)^{-2}\Big)$$

$$\leq \sup_{w \geq N} \sup_{0 \leq x \leq 1} \left[\mathbf{P}\Big(\frac{|x + S_{wt}|}{(Vw)^{1/2}} \leq 1 : 0 \leq t \leq L(1+2\epsilon)^{-2}\Big)\right]^{M(n)-1}$$

where the rescaling factor $w = (1+2\epsilon)^2 V^{-1} d_k^2$ is at least N by the previous choice of m_0. Taking the log, dividing by $L(1+2\epsilon)^{-2}$ and applying (3) gives

$$L^{-1}(1+2\epsilon)^2 \ln \mathbf{P}(f(k) \leq S_k \leq g(k) : k \leq n) \leq (M(n)-1)(-\frac{\pi^2}{8} + \epsilon)$$

by choice of N. Plugging in the asymptotic value of $M(n)$ from the claim above gives that for sufficiently large n,

$$\ln \mathbf{P}(f(k) \leq S_k \leq g(k) : k \leq n)$$

$$\leq -\frac{\pi^2}{8}(M(n)-1)L(1+2\epsilon)^{-2}$$

$$\leq \epsilon + 4(-\frac{\pi^2}{8} + \epsilon)V(1+2\epsilon)^{-1} \sum_{k=1}^{n} (f(k) - g(k))^{-2}$$

for large n. Letting $\epsilon \to 0$ proves that the limsup in (5) is at most $-\frac{\pi^2}{8}V$.

Proving that the liminf is at least $-\frac{\pi^2}{8}V$ is almost identical. Fixing $L, \epsilon > 0$, choose N and m_0 as before and this time define R_k to lie between f and g instead of covering them:

$$R_k = \{(x,y) : m_k \leq x \leq m_{k+1} \text{ and } y_k - (1-2\epsilon)d_k \leq y \leq y_k + (1-2\epsilon)d_k\}.$$

Let G_k be the event that $(i, S_i) \in R_k$ for $m_k \leq i \leq m_{k+1}$ and that $y_k - \frac{1}{4}d_k \leq S_{m_{k+1}} \leq y_k + \frac{1}{4}d_k$. Since $m_{M(n)+1} \geq n$, the probability we are trying to bound from below is at least $\mathbf{P}(g(k) \leq S_k \leq f(k) : k = 1, \ldots, m_{M(n)+1})$, which may be written as

$$\mathbf{P}(g(k) \leq S_k \leq f(k) : k = 1, \ldots, m_0) \prod_{i=0}^{M(n)} \mathbf{P}(G_k \,|\, S_{m_k}, G_{k-1}).$$

When G_{k-1} occurs, the value of S_k is certainly between $y_k - \frac{1}{2}d_k$ and $y_k + \frac{1}{2}d_k$, so

$$\mathbf{P}(G_k \mid S_{m_k}, G_{k-1}) \geq \inf_{y_k - \frac{1}{2}d_k \leq x \leq y_k + \frac{1}{2}d_k} \mathbf{P}(G_k \mid S_k = x).$$

Now rescaling each rectangle, applying (4) with $c_1 = -(1/2)(1-2\epsilon)^{-2}, c_2 = (1/2)(1-2\epsilon)^{-2}, c_3 = -1/4, c_4 = 1/4$, and taking limits establishes that the liminf in (5) is at least $-\frac{\pi^2}{8}V$, finishing the proof of the lemma.

Finally, the corollary is proved by letting $f(n) = c_2 n^{1/3}, g(n) = c_1 n^{1/3}$, and verifying the flatness hypothesis; summing $[f(k) - g(k)]^{-2}$ from 1 to n gives $(3 + o(1))(c_2 - c_1)^{-2} n^{1/3}$ and the desired conclusion follows. □

The next step is to apply these random walk estimates to prove a tree-indexed version of Kesten's result on branching Brownian motion staying above zero.

Lemma 8 *Let $\Gamma, X_\sigma, S_\sigma$ and β be as above. Assume that X is top-heavy and let λ_0 be the value of λ minimizing $\mathbf{E}e^{\lambda X}$ (which must exist and be strictly less than one, according to the definition of the term top-heavy). There exists a positive real number c, independent of Γ, such that if $|\Gamma_n| \leq e^{cn^{1/3} + n\beta}$ for all n, then*

$$\mathbf{P}(\max_{\sigma \in \Gamma_n} \min_{\tau \leq \sigma} S(\tau) \geq -2c(1 - \lambda_0)^{-1} n^{1/3}) \to 0 \tag{7}$$

as $n \to \infty$. In other words with high probability, for each sufficiently large n, no path from ρ of length n stays above $-2c(1 - \lambda_0)^{-1} n^{1/3}$.

To establish this, first record some elementary facts about large deviations.

Proposition 9 *Let X_n be i.i.d. and $S_k = \sum_{i=1}^k X_i$. Let $\beta(X)$ be the backwards push, and let $\lambda_0(X)$ be the $\lambda \in [0, 1]$ minimizing $\mathbf{E}e^{\lambda X}$. Then the following three inequalities hold.*

(i) *For any real u,*

$$\mathbf{P}(S_n \geq u) \leq e^{-\beta n - \lambda_0 u};$$

(ii) *For any real y,*

$$\mathbf{E}e^{S_n} I(S_n \leq y) \leq (1 - \lambda_0)^{-1} e^{(1 - \lambda_0)y - \beta n}.$$

(iii) For any real y,

$$\mathbf{E}e^{S_n \wedge y} \leq [1 + (1-\lambda_0)^{-1}]e^{(1-\lambda_0)y - \beta n}.$$

Proof: The first claim is just Markov's inequality:

$$\mathbf{P}(S_n \geq u) \leq e^{-\lambda_0 u}\mathbf{E}e^{-\lambda_0 S_n}.$$

For the second claim, integrate the first by parts:

$$\begin{aligned}
\mathbf{E}(e^{S_n} I(S_n \leq y)) &= \int_{-\infty}^{y} e^u \mathbf{P}(S_n \in du) \\
&= \int_{-\infty}^{y} e^u \mathbf{P}(u \leq S_n \leq y) du \\
&\leq \int_{-\infty}^{y} e^u \mathbf{P}(u \leq S_n) du \\
&\leq \int_{-\infty}^{y} e^u e^{-\lambda_0 u - \beta n} du \\
&= (1-\lambda_0)^{-1} e^{(1-\lambda_0)y - \beta n}.
\end{aligned}$$

Finally, the third claim follows from the first two, using

$$\mathbf{E}e^{S_n \wedge y} = \mathbf{E}e^{S_n} I(S_n \leq y) + e^y \mathbf{P}(S_n > y).$$

It should be remarked that more careful estimates give an extra factor of $(1 + o(1))(2\pi n \mathbf{E}Y^2)^{-1/2}$ in the RHS of each inequality which is then asymptotically sharp. □

Next, plug this into a first moment calculation to establish:

Proposition 10 *Let $\Gamma, X_\sigma, S_\sigma$ and β be as above. Suppose that $|\Gamma_n| \leq e^{cn^{1/3} + n\beta}$ for some $c > 0$. Then for any $\epsilon > 0$,*

$$\mathbf{P}(S(\sigma) \geq (1+\epsilon)c\lambda_0^{-1}n^{1/3} \textit{ for some } \sigma \textit{ with } |\sigma| \leq n) \to 0$$

as $n \to \infty$.

Proof: For each fixed L it is clear that

$$\mathbf{P}(S(\sigma) \geq (1+\epsilon)c\lambda_0^{-1}n^{1/3} \text{ for some } \sigma \text{ with } |\sigma| \leq L) \quad (8)$$

goes to zero as $n \to \infty$. On the other hand,

$$\begin{aligned} &\mathbf{P}(S(\sigma) \geq (1+\epsilon)c\lambda_0^{-1}n^{1/3} \text{ for some } \sigma \text{ with } n \geq |\sigma| > L) \\ \leq\ &\mathbf{P}(S(\sigma) \geq (1+\epsilon)c\lambda_0^{-1}|\sigma|^{1/3} \text{ for some } \sigma \text{ with } n \geq |\sigma| > L) \\ \leq\ &\sum_{m>L} \mathbf{P}(S(\sigma) \geq (1+\epsilon)c\lambda_0^{-1}m^{1/3} \text{ for some } \sigma \in \Gamma_m). \end{aligned}$$

For $\sigma \in \Gamma_m$, Proposition 9 part (i) implies

$$\mathbf{P}(S(\sigma) \geq (1+\epsilon)c\lambda_0^{-1}m^{1/3}) \leq e^{-\lambda_0(1+\epsilon)c\lambda_0^{-1}m^{1/3}-\beta m}.$$

Multiplying by $|\Gamma_m|$ gives

$$\mathbf{P}(S(\sigma) \geq (1+\epsilon)c\lambda_0^{-1}m^{1/3} \text{ for some } \sigma \in \Gamma_m) \leq e^{-\epsilon c n^{1/3}}.$$

This is summable in m, so the sum over $m > L$ goes to zero as $L \to \infty$, which together with (8) proves the proposition. □

Proof of Lemma 8: Let μ be the common distribution of the $X(\sigma)$ and let $Y_1, \ldots, Y_n$ be i.i.d. random variables whose law μ' satisfies

$$\frac{d\mu'}{d\mu}(x) = e^{\lambda_0 x}/\mathbf{E}e^{\lambda_0 X} = e^{\lambda_0 x + \beta(\mu)}. \tag{9}$$

Informally, μ' is μ tilted in the large deviation sense so as to have mean zero. The assumption that μ is top-heavy by definition implies that $\mathbf{E}e^{\lambda X} < \infty$ for λ in some neighborhood of λ_0, hence $\mathbf{E}e^{\lambda Y_1} < \infty$ for λ in some neighborhood of zero, and in particular $V \stackrel{def}{=} \mathbf{E}Y_1^2 < \infty$.

Choose a positive real c for which

$$c + 2\lambda_0(1-\lambda_0)^{-1}c - \frac{\pi^2}{8}\frac{3V}{(2c\lambda_0^{-1} + 2c(1-\lambda_0)^{-1})^2} < 0. \tag{10}$$

Let

$$A_n = \left\{ \max_{\sigma \in \Gamma_n} \min_{\tau \leq \sigma} S(\tau) \geq -2c(1-\lambda_0)^{-1}n^{1/3} \right\}$$

be the event in (7). Let G_n be the event

$$\bigcup_{|\sigma| \leq n} \{S(\sigma) \geq 2c\lambda_0^{-1}n^{1/3}\}$$

and let $H_n = A_n \setminus G_n$. Proposition 10 shows that $\mathbf{P}(G_n) \to 0$ so to show that $\mathbf{P}(A_n) \to 0$ it suffices to show that $\mathbf{P}(H_n) \to 0$.

To see this, fix $\sigma \in \Gamma_n$ and write $\mathbf{P}(H_n) \leq |\Gamma_n| Q_n$ where

$$Q_n = \mathbf{P}(2c\lambda_0^{-1} n^{1/3} \geq S(\tau) \geq -2c(1-\lambda_0)^{-1} n^{1/3} \text{ for all } \tau \leq \sigma).$$

Let ν be the law in $\mathbf{R}^n$ of the sequence $(S(\sigma_1), \ldots, S(\sigma_n))$, where $\sigma_1, \ldots, \sigma_n$ is the path from $\rho = \sigma_0$ to $\sigma = \sigma_n$; of course ν is just the law in $\mathbf{R}_n$ of a random walk whose steps have law μ. Recalling the tilted variables Y_n, write $T_n = \sum_{i=1}^n Y_i$. Let ν' denote the law in $\mathbf{R}_n$ of $(T_1, \ldots, T_n)$ and observe that

$$\frac{d\nu'}{d\nu}(s_1, \ldots, s_n) = e^{\lambda_0 s_n} / \mathbf{E} e^{\lambda_0 S_n} = e^{\lambda_0 s_n + n\beta}.$$

Use this to get an upper bound on Q_n as follows.

$$\begin{aligned}
& Q_n \\
= & \int I(2c\lambda_0^{-1} n^{1/3} \geq s_i \geq -2c(1-\lambda_0)^{-1} n^{1/3} \text{ for all } i \leq n) \\
& d\nu(s_1, \ldots, s_n) \\
= & \int I(2c\lambda_0^{-1} n^{1/3} \geq s_i \geq -2c(1-\lambda_0)^{-1} n^{1/3} \text{ for all } i \leq n) \\
& \qquad \frac{d\nu}{d\nu'}(s_1, \ldots, s_n) d\nu'(s_1, \ldots, s_n) \\
\leq & \int I(2c\lambda_0^{-1} n^{1/3} \geq s_i \geq -2c(1-\lambda_0)^{-1} n^{1/3} \text{ for all } i \leq n) \\
& \qquad \left[\sup_{s_n \geq -2c(1-\lambda_0)^{-1} n^{1/3}} \frac{d\nu}{d\nu'}(s_1, \ldots, s_n) \right] d\nu(s_1, \ldots, s_n) \\
= & \int I(2c\lambda_0^{-1} n^{1/3} \geq s_i \geq -2c(1-\lambda_0)^{-1} n^{1/3} \text{ for all } i \leq n) \\
& \qquad \exp(2c\lambda_0 (1-\lambda_0)^{-1} n^{1/3} - n\beta) \, d\nu(s_1, \ldots, s_n) \\
\leq & \exp(2c\lambda_0(1-\lambda_0)^{-1} n^{1/3} - n\beta) \\
& \qquad \mathbf{P}(2c\lambda_0^{-1} n^{1/3} \geq T_i \geq -2c(1-\lambda_0)^{-1} \text{ for all } i \leq n) \\
\leq & \exp\left(-n\beta + n^{1/3} [\frac{2c\lambda_0}{1-\lambda_0} - \frac{\pi^2}{8} \frac{3V}{(2c\lambda_0^{-1} + 2c(1-\lambda_0)^{-1})^2} + o(1)] \right)
\end{aligned}$$

by Corollary 7. Thus $\mathbf{P}(H_n) \leq |\Gamma_n| Q_n \leq \exp(cn^{1/3} + n\beta) Q_n \leq$

$$\exp\left(n^{1/3} \left[c + 2c\lambda_0(1-\lambda_0)^{-1} - \frac{\pi^2}{8} \frac{3V}{(2c\lambda_0^{-1} + 2c(1-\lambda_0)^{-1})^2} + o(1) \right] \right).$$

By choice of c, this is $\exp((K+o(1))n^{1/3})$ for some $K<0$, so $\mathbf{P}(H_n)\to 0$, proving the lemma. □

4 Proof of the main theorem

The case where $X \equiv -\beta$ is done in [5]. For part (i) of the nondegenerate case, use Lemma 2, showing that $\sum_{|\sigma|=n} U(\sigma)$ goes to zero in probability by computing a truncated expectation. Let c_1 be the constant c from Lemma 8 and let $U_n = \sum_{|\sigma|=n} U(\sigma)$. Let G_n be the event that $\max_{|\sigma|=n} U(\sigma) \geq \exp(-2c_1(1-\lambda_0)^{-1}n^{1/3})$. Then

$$\begin{aligned} P(U_n > \epsilon) &\leq \mathbf{P}(G_n) + \mathbf{P}(U_n > \epsilon \text{ and } G_n^c) \\ &\leq \mathbf{P}(G_n) + \epsilon^{-1}\mathbf{E}U_n I(G_n^c). \end{aligned}$$

Lemma 8 showed that $\mathbf{P}(G_n) \to 0$, so it remains to show that for any ϵ, $\mathbf{E}(U_n I(G_n^c)) \to 0$.

Observe that $U_\sigma I(G_n^c) \leq \exp(S_\sigma \wedge -2c(1-\lambda_0)^{-1}n^{1/3})$. Hence for $\sigma \in \Gamma_n$,

$$\begin{aligned} &\mathbf{E}U_n I(G_n^c) \\ = \;& |\Gamma_n|\mathbf{E}U_\sigma I(G_n^c) \\ \leq \;& |\Gamma_n|\mathbf{E}\exp(S_\sigma \wedge -2c_1(1-\lambda_0)^{-1}n^{1/3}) \\ \leq \;& |\Gamma|[1+(1-\lambda_0)^{-1}]\exp((1-\lambda_0)(-2c_1(1-\lambda_0)^{-1}n^{1/3}) - n\beta) \\ & \quad \text{by Proposition 9, part } (iii), \\ \leq \;& [1+(1-\lambda_0)^{-1}]\exp(-c_1 n^{1/3}) \end{aligned}$$

by the assumption on $|\Gamma_n|$. This goes to zero, thus $U_n \to 0$ in probability, proving part (i) of Theorem 1.

Part (ii) is proved by exhibiting an infinite path along which the resistances e^{-S_σ} are summable. In fact the proof finds an infinite path along which $S(\sigma)/c|\sigma|^{1/3}$ is bounded above and below.

Pick any $c, L, \epsilon > 0$ and any $c_2 = K + 2c\lambda_0 + (\frac{\pi^2}{8})\frac{3V}{(.9c)^2}$, where M shall be chosen later. Define W to be the random subset of vertices σ of Γ with the property that for every $\tau \leq \sigma$ with $|\tau| > L$,

$$\frac{c|\tau|^{1/3}}{10} \leq S_\tau \leq c|\tau|^{1/3},$$

where L is large enough so that W intersects each Γ_n with positive probability. To prove the theorem, it suffices to show that W is infinite with positive probability; this follows from Lemma 3 and the hypothesis of the theorem, provided that

$$\sup_n a(n,k)/a(n,n)^2 \leq e^{c_2 k^{1/3} + k\beta} \tag{11}$$

for all but finitely many k, where $a(n,k) = \mathbf{P}(\rho \leftrightarrow \sigma, \tau)$ for vertices $\sigma, \tau \in \Gamma_n$ with $|\sigma \wedge \tau| = k$.

To establish (11), begin with

$$a(n,k)/a(n,n)^2 = \mathbf{P}(\sigma \wedge \tau \in W)^{-1} \frac{\mathbf{P}(\sigma \in W \mid \tau \in W)}{\mathbf{P}(\sigma \in W \mid \sigma \wedge \tau \in W)}. \tag{12}$$

Recall the tilted random variables Y_n and T_n whose law μ' is defined by (9).

Fix any $k \leq n$ and $\sigma, \tau \in \Gamma_n$ with $|\sigma \wedge \tau| = k$. Let $C(a,b)$ denote the set of sequences $(s_a, \ldots, s_b) \in \mathbf{R}^{b-a+1}$ for which $cs_j^{1/3}/10 \leq s_j \leq cs_j^{1/3}$ for all $j \in [a,b]$. Let ν and ν' respectively denote the law of $(S_1, \ldots, S_k)$ and $(T_1, \ldots, T_k)$ and let ν_y and ν'_y denote the laws of $(S_k, \ldots, S_n)$ and $(T_k, \ldots, T_n)$ conditioned respectively on $S_k = y$ and $T_k = y$. Write the first factor on the RHS of (12) as

$$\left[\int I(\alpha \in C(1,k))\, d\nu(\alpha)\right]^{-1}.$$

Changing the integrating measure to ν' yields

$$\begin{aligned} &\left[\int I(\alpha \in C(1,k))\, d\nu'(\alpha) \frac{d\nu(\alpha)}{d\nu'(\alpha)}\right]^{-1} \\ \leq\ & \sup_{\alpha \in C(1,k)} \frac{d\nu'(\alpha)}{d\nu(\alpha)} \left[\int I(\alpha \in C(1,k))\, d\nu'(\alpha)\right]^{-1} \\ =\ & e^{\lambda_0 c k^{1/3} + k\beta} (\nu'(C(1,k)))^{-1} \\ =\ & \exp\left[\lambda_0 c k^{1/3} + k\beta + k^{1/3}(\frac{\pi^2}{8}\frac{3V}{(.9c)^2} + o(1))\right] \end{aligned} \tag{13}$$

by Corollary 7, since ν' is the law of a mean zero, finite variance random walk, and it has been assumed that $\nu'(C(1,k))$ never vanishes.

For the second factor on the RHS of (12), let μ_1 be the law of $S(\sigma \wedge \tau)$ conditional on $\tau \in W$ and let μ_2 be the law of $S(\sigma \wedge \tau)$ conditional on $\sigma \wedge \tau \in W$. The second factor is then

$$\frac{\int \left[\int I(\alpha \in C(k,n))\, d\nu_y(\alpha)\right] d\mu_1(y)}{\int \left[\int I(\alpha \in C(k,n))\, d\nu_y(\alpha)\right] d\mu_2(y)}. \tag{14}$$

Changing the integrating measure again, this becomes

$$\frac{\int\int I(\alpha \in C(k,n))\, \frac{d\nu}{d\nu'}(\alpha)\, d\nu'_y(\alpha) d\mu_1(y)}{\int\int I(\alpha \in C(k,n))\, \frac{d\nu}{d\nu'}(\alpha)\, d\nu'_y(\alpha) d\mu_2(y)}$$

$$\leq \frac{\int I(\alpha \in C(k,n)) d\nu'_y(\alpha) d\mu_1(y)}{\int I(\alpha \in C(k,n)) d\nu'_y(\alpha) d\mu_2(y)} \frac{\sup_{\alpha \in C(k,n)} \frac{d\nu}{d\nu'}(\alpha)}{\inf_{\alpha \in C(k,n)} \frac{d\nu}{d\nu'}(\alpha)}$$

$$\leq \frac{\sup_y \nu'_y(C(k,n))}{\inf_y \nu'_y(C(k,n))} \frac{\sup_{\alpha \in C(k,n)} \frac{d\nu}{d\nu'}(\alpha)}{\inf_{\alpha \in C(k,n)} \frac{d\nu}{d\nu'}(\alpha)}$$

$$= \frac{\sup_y \nu'_y(C(k,n))}{\inf_y \nu'_y(C(k,n))}\, e^{\lambda_0 c k^{1/3}}.$$

The argument is then finished by establishing

$$\frac{\sup_y \nu'_y(C(k,n))}{\inf_y \nu'_y(C(k,n))} \leq e^{Mk^{1/3}} \tag{15}$$

for some $M > 0$, since then multiplying inequalities (13) and (15) bounds $a(n,k)/a(n,n)^2$ from above by $\exp\left[2\lambda_0 c k^{1/3} + k\beta + k^{1/3}(\frac{\pi^2}{8}\frac{3V}{(.9c)^2} + M)\right]$, which is at most $e^{c_2 k^{1/3} + k\beta}$ by choice of c_2, yielding (11).

It remains to establish (15). An argument is given for the case where the distribution of the $X(\sigma)$'s has an absolutely continuous component near zero, the lattice case being similar. By hypothesis, the measure μ' has density at most A and is greater than some constant, $a > 0$, times Lebesgue measure on some interval $(-b,b)$. Call this latter measure π. Let $l = k + ck^{1/3}/b$. Write

$$\nu'_y(C(k,n)) = \int m_z(C(k+l,n))\, dm^y(z)$$

where m_z is the law of $(T_{k+l}, \ldots, T_n)$ conditioned on $T_{k+l} = z$ and m^y is the (deficient) law of T_{k+l} conditioned on $T_k = y$ and killed if $T_i \notin [ci^{1/3}/10, ci^{1/3}]$ for some $k \leq i \leq k+l$. It is easy to see that

$$m^y \geq \pi^y \geq a^l \text{ times Lebesgue measure on } [c(k+l)^{1/3}/10, c(k+l)^{1/3}]$$

where π^y is the measure on sequences $s_k, \ldots, s_{k+l}$ with $s_k = y$, having increments distributed as π and killed if $s_i \notin [ci^{1/3}/10, ci^{1/3}]$ for some $k \leq i \leq k+l$. Then for any $x, y \in [ck^{1/3}/10, ck^{1/3}]$,

$$\begin{aligned} & \nu_y'(C(k,n)) \\ = & \int m_z(C(k+l,n))\, dm^y(z) \\ \geq & a^l \int m_z(C(k+l,n))\, d\lambda(z) \\ \geq & (a/A)^l \int m_z(C(k+l,n))\, dm_x(z) \\ = & (a/A)^l \nu_x'(C(k,n)). \end{aligned}$$

Checking against the value of l proves (15) and the theorem. □

References

[1] Benjamini, I. and Peres, Y. (1993). Markov chains indexed by a tree. *Ann. Probab. to appear.*

[2] Doyle, P. and Snell, J. L. (1984). Random walks and electrical networks. Mathematical Association of America: Washington.

[3] Durrett, R. (1984). Brownian motion and martingales in analysis. Wadsworth: Monterey, CA.

[4] Kesten, H. (1978). Branching Brownian motion with absorption. *Stoch. Pro. Appl.* **7** 9 - 47.

[5] Lyons, R. (1990). Random walks and percolation on trees. *Ann. Probab.* **18** 931 - 958.

[6] Lyons, R. (1992). Random walks, capacity and percolation on trees. *Ann. Probab. to appear.*

[7] Lyons, R. and Pemantle, R. (1992). Random walk in a random environment and first-passage percolation on trees. *Ann. Probab.* **20** 125 - 136.

[8] Pemantle, R. (1988). Phase transition in Reinforced random walk and RWRE on trees. *Ann. Probab.* **16** 1229 - 1241.

[9] Pemantle, R. and Peres, Y. (1991). Critical random walk in random environment on a tree. *Preprint.*

Keywords: RWRE, critical RWRE, tree, tree-indexed random walk
Subject classification: Primary: 60J15. Secondary: 60G60, 60G70, 60E07.

Uniform Invariance Principles for Intersection Local Times

Jay Rosen

1 Introduction

Let S_n be a strongly aperiodic stable random walk, i.e. in the domain of attraction of a non-degenerate stable random variable U of index β in $\mathbf{R}^d$. Thus

$$\frac{S_n}{b_n} \stackrel{\mathcal{L}}{\Longrightarrow} U$$

for some b_n which is regularly varying of order β.

Given k independent copies $S_n^{(1)}, \ldots, S_n^{(k)}$ of S_n we define their k-fold intersection local time by

$$(1.1) \quad I_k(x,t) = \sum_{i_1,\ldots,i_k=1}^{t} \delta(S_{i_2}^{(2)} - S_{i_1}^{(1)}, x_1) \cdots \delta(S_{i_k}^{(k)} - S_{i_{k-1}}^{(k-1)}, x_{k-1})$$

where

$$\delta(i,j) = \begin{cases} 1 & \text{if } i = j \\ 0 & \text{otherwise} \end{cases}$$

is the usual Kronecker delta function and $x = (x_1, \ldots, x_{k-1}) \in (\mathbf{Z}^d)^{k-1}$, $t \in \mathbf{Z}_+$. This definition is extended to $x \in (\mathbf{R}^d)^{k-1}$, $t \in \mathbf{R}_+$ by linear interpolation.

Let Z_t denote the stable Levy process of index β in $\mathbf{R}^d$ with $Z_1 = U$. We use $p_t(x)$ for the transition density of Z_t. If $Z_t^{(1)}, \ldots, Z_t^{(k)}$ denote k independent copies of Z_t, we set

$$\alpha_k(\epsilon, x, t) = \int_{D_k} p_\epsilon(Z_{s_2}^{(2)} - Z_{s_1}^{(1)} - x_1) \cdots p_\epsilon(Z_{s_k}^{(k)} - Z_{s_{k-1}}^{(k-1)} - x_{k-1})\, ds_1 \ldots ds_k$$

where $D_k = \{(s_1, \ldots, s_k) \in \mathbf{R}^k \mid 0 \le s_1 \le \ldots \le s_k \le t\}$. It is known that if $\beta > d - d/k$ then $\alpha_k(\epsilon, x, t)$ converges, as $\epsilon \to 0$, to a random variable, called the k-fold intersection local time, and denoted $\alpha_k(x,t)$. Convergence is locally uniform both a.s. and in all
L^p spaces. The k-fold intersection local time $\alpha_k(x,t)$ is jointly continuous in (x,t).

Theorem 1 *If $\beta > d - d/k$ then*

$$I_k(b_n x, nt)\, b_n^{(k-1)d}/n^k \Longrightarrow \alpha_k(x,t)$$

as $n \to \infty$, where we have weak convergence of processes in $C(\mathbf{R}^{d(k-1)} \times \mathbf{R}_+)$.

Such a theorem is referred to as a uniform invariance principle for intersection local times. It "uniformizes" our work in [5] where the convergence in Theorem 1 is proven for fixed x and t. Our present theorem was inspired by the work of B ass and Khoshnevisan [1, 2] who establish Theorem 1 for random walks with finite variance, in which case $\beta = 2$ and $b_n = \sqrt{n}$. Their work, in turn, was motivated by the uniform invariance principles of Perki ns [4] and Borodin [3] for ordinary local times. We should also mention that in [2], Bass and Khoshnevisan obtain a strong invariance principle for intersection local times of certain random walks. More preci sely, they show that if a random walk converges almost surely to Brownian motion at a certain rate, then this will also hold for their intersection local times.

2 Proof of Theorem 1

The proof of a uniform invariance principle consists of two parts: a proof that the finite dimensional distributions converge, and a proof of tightness. The proof that the finite dimensional distributions converge proceeds almost exactly as in [5] where convergence of the marginal distributions is established. We shall only recall the basic ideas and mention the necessary modifications.

By a change of variables, for $x \in (\mathbf{Z}^d)^{k-1}/b_n$, $t \in \mathbf{Z}_+/n$ we have

$$\begin{aligned} (2.1)\qquad L_k(n,x,t) &\stackrel{\text{def}}{=} I_k(b_n x, nt)\, b_n^{(k-1)d}/n^k \\ &= \sum_{i_1,\dots,i_k=1}^{nt} \frac{1}{n^k} \prod_{j=2}^{k} \frac{1}{(2\pi)^d} \int_{|p_j|_0 \le \pi b_n} \exp(ip_j \frac{S^{(j)}_{i_j} - S^{(j-1)}_{i_{j-1}} - b_n x_{j-1}}{b_n})\, dp_j \end{aligned}$$

where for a vector $y = (y_1, \dots, y_d) \in \mathbf{R}^d$ we use $|y|_0 \stackrel{\text{def}}{=} \max_i |y_i|$.

We then define a 'link'

$$\begin{aligned} (2.2)\quad L_k(\epsilon, n, x, t) \stackrel{\text{def}}{=} &\sum_{i_1,\dots,i_k=1}^{nt} \frac{1}{n^k} \prod_{j=2}^{k} \frac{1}{(2\pi)^d} \int_{|p_j|_0 \le \pi b_n} \\ &\exp(ip_j \frac{S^{(j)}_{i_j} - S^{(j-1)}_{i_{j-1}} - b_n x_{j-1}}{b_n}) \Phi_\epsilon(p_j)\, dp_j \end{aligned}$$

where $\Phi_\epsilon(p)$ is the characteristic function of Z_ϵ, i.e. the Fourier transform of $p_\epsilon(x)$.

In lemma 1 of [5] we essentially prove that

$$||L_k(n,x,t) - L_k(\epsilon,n,x,t)||_2 \leq c\epsilon^\gamma$$

for some $c < \infty, \gamma > 0$ uniformly in $(x,t) \in \mathbf{R}^{d(k-1)} \times [0,T]$ for any $T < \infty$. Hence for any fixed $\lambda_i, z_i, t_i \quad i = 1,\ldots,m$, if we set

$$L(\epsilon,n) \stackrel{\text{def}}{=} \sum_{i=1}^{m} \lambda_i L_k(\epsilon,n,z_i,t_i)$$

and

$$L(n) \stackrel{\text{def}}{=} \sum_{i=1}^{m} \lambda_i L_k(n,z_i,t_i)$$

we have that

$$|E(e^{iL(n)}) - E(e^{iL(\epsilon,n)})| \leq c\epsilon^\gamma. \tag{2.3}$$

On the other hand, it follows from the locally uniform convergence of $\alpha_k(\epsilon,x,t)$ to $\alpha_k(x,t)$ that if we set

$$\alpha(\epsilon) \stackrel{\text{def}}{=} \sum_{i=1}^{m} \lambda_i \alpha_k(\epsilon,z_i,t_i)$$

and

$$\alpha \stackrel{\text{def}}{=} \sum_{i=1}^{m} \lambda_i \alpha_k(z_i,t_i)$$

we can choose $\epsilon_0 > 0$ such that for any given $\delta > 0$ we have both $c\epsilon_0^\gamma \leq \delta$ and

$$|E(e^{i\alpha}) - E(e^{i\alpha(\epsilon_0)})| \leq \delta. \tag{2.4}$$

From (2.2) we see that

$$\begin{aligned} (2.5) \quad & L_k(\epsilon,n,x,t) \\ &= \sum_{i_1,\ldots,i_k=1}^{nt} \frac{1}{n^k} \prod_{j=2}^{k} p_{\epsilon_0}\Big(\frac{S_{i_j}^{(j)} - S_{i_{j-1}}^{(j-1)} - b_n x_{j-1}}{b_n}\Big)\, dp_j + O(e^{-\epsilon_0 n^{1/2}}). \end{aligned}$$

We see from this, using [6], that we can find n_0 such that for all $n \geq n_0$ we have

$$|E(e^{iL(\epsilon_0,n)}) - E(e^{i\alpha(\epsilon_0)})| \leq \delta.$$

Together with (2.4) and (2.5) this shows the convergence of finite dimensional distributions.

To prove tightness it suffices to show that we can find some $\gamma > 0$ such that for any even m

$$E\{(L_k(n,x,t) - L_k(n,x',t'))^m\} \leq c|(x,t) - (x',t')|^{m\gamma} \tag{2.6}$$

uniformly over $n \in \mathbf{Z}_+$, $x, x' \in \mathbf{R}^d$ and $t, t' \in [0,T]$.

We begin by showing how to get a bound on

$$E\{(L_k(n,x,t))^m\}$$

which is uniform in $n \in \mathbf{Z}_+$, $x \in \mathbf{R}^d$ and $t \in [0,T]$. Once we see how to accomplish this, it will be easy to establish (2.6).

Recalling (2.1) we have

$$\begin{aligned}
(2.7)\qquad & E\{(L_k(n,x,t))^m)\} \\
&= E\{\prod_{h=1}^m \sum_{i_{1,h},\ldots,i_{k,h}=1}^{nt} \frac{1}{n^k} \prod_{j=2}^k \frac{1}{(2\pi)^d} \\
&\qquad \int_{|p_{j,h}|_0 \le \pi b_n} \exp(ip_{j,h} \frac{S^{(j)}_{i_{j,h}} - S^{(j-1)}_{i_{j-1,h}} - b_n x_{j-1}}{b_n})\, dp_{j,h}\} \\
&= \frac{1}{(2\pi)^{d(k-1)m}} \int_{|p_{j,h}| \le \pi b_n} F(p,x) \prod_{j=1}^k \frac{1}{n^{km}} \\
&\qquad \sum_{i_{j,1},\ldots,i_{j,m}=1}^{nt} E\{\exp(i\sum_{h=1}^m \frac{(p_{j,h} - p_{j+1,h})}{b_n} S^{(j)}_{i_{j,h}})\}\, dp
\end{aligned}$$

where

$$F(p,x) = \exp(i\sum_{h=1}^m \sum_{j=2}^k p_{j,h} x_{j-1})$$

and we have set $p_{1,h} = p_{k+1,h} = 0$.

Let $\pi^1, \ldots, \pi^k$ be k not necessarily distinct permutations of $\{1,\ldots,m\}$ and let

$$\Delta(\pi^1,\ldots,\pi^k) = \{i_{j,h} \,|\, i_{j,\pi^j_l} \le i_{j,\pi^j_{l+1}}\,; j = 1,\ldots,k;\, l = 1,\ldots,m\}$$

and note that on $\Delta(\pi^1,\ldots,\pi^k)$ we have

$$\begin{aligned}
(2.8)\qquad & E\{\exp(i\sum_{h=1}^m \frac{(p_{j,h} - p_{j+1,h})}{b_n} S^{(j)}_{i_{j,h}})\} \\
&= E\{\exp(i\sum_{h=1}^m u_{j,h}(S^{(j)}_{i_{j,\pi^j_h}} - S^{(j)}_{i_{j,\pi^j_{h-1}}})/b_n)\} \\
&= \prod_{h=1}^m \varphi^{\pi^j_h - \pi^j_{h-1}}(u_{j,h}/b_n)
\end{aligned}$$

where

$$\varphi(u) = E(\exp(iuS_1))$$

and

$$u_{j,h} = \sum_{l=1}^{h}(p_{j,\pi_l^j} - p_{j+1,\pi_l^j}).$$

Note that

$$\text{(2.9)} \quad span\{u_{j,h} \,|\, h = 1, \ldots, m\} = span\{p_{j,h} - p_{j+1,h} \,|\, h = 1, \ldots, m\}$$

and that if $\tilde{\pi}^j$ denotes the inverse of the permutation π^j we have

$$\text{(2.10)} \qquad p_{j,h} - p_{j+1,h} = u_{j,\tilde{\pi}_h^j} - u_{j,\tilde{\pi}_h^j - 1}$$

If we define

$$G_r(n,u) = \frac{1}{n}\sum_{j=0}^{nT} |\varphi(u/b_n)|^{jr}$$

then it is clear from the above that (2.7) can be bounded by a sum over regions $\Delta(\pi^1, \ldots, \pi^k)$ of integrals of the form

$$\text{(2.11)} \quad \int_{|p_{j,h}| \le \pi b_n;\, j=2,\ldots,k} \prod_{j=1}^{k}\prod_{h=1}^{m} G_1(n, u_{j,h})\, dp$$

$$\le \prod_{i=1}^{k}\{\int_{|p_{j,h}| \le \pi b_n;\, j=2,\ldots,k} \prod_{j=1; j\neq i}^{k}\prod_{h=1}^{m} G_1^{k/(k-1)}(n, u_{j,h})\, dp\}^{(k-1)/k}.$$

By [5] we know that for any $\epsilon, r > 0$ and $T < \infty$ we have

$$G_r(n,u) \le \frac{c}{1 + ((u))^{\beta - \epsilon}}$$

where $((u))$ denotes the norm of the smallest vector which equals u mod 2π. Since, by (2.9), for any i

$$span\{u_{j,h} \,|\, h = 1, \ldots, m;\ j = 1, \ldots, k;\ j \neq i\}$$
$$= span\{p_{j,h} \,|\, h = 1, \ldots, m;\ j = 2, \ldots, k\}$$

we have that (2.11) and hence (2.7) is bounded uniformly if $k/(k-1)\beta > d$.

We now show how to modify the above estimates to get (2.6). First , we have

$$\text{(2.12)} E\{(L_k(n,x,t) - L_k(n,x',t'))^m\}$$
$$\le E\{(L_k(n,x,t) - L_k(n,x',t))^m\} + E\{(L_k(n,x',t) - L_k(n,x',t'))^m\}$$

and we can handle the x and t variation seperately.

The first term in (2.12), the x variation, can be written as in (2.7) except that the factor $F(p,x)$ will be replaced by

$$H(p,x,x') = \prod_{h=1}^{m}\{\exp(i\sum_{j=2}^{k} p_{j,h}x_{j-1}) - \exp(i\sum_{j=2}^{k} p_{j,h}x'_{j-1})\}.$$

Since for any $0 \leq \delta \leq 1$ we have

$$|H(p,x,x')| \leq c \prod_{h=1}^{m} (\sum_{j=2}^{k} |p_{j,h}|)^{\delta} |x - x'|^{\delta}$$

and using (2.10) for any $i = 1, \ldots, k$ we have

$$\begin{aligned}
&\prod_{h=1}^{m} \sum_{j=2}^{k} |p_{j,h}| \\
&\leq c \prod_{h=1}^{m} \sum_{j=1; j \neq i}^{k} |p_{j,h} - p_{j-1,h}| \\
&\leq c \prod_{h=1}^{m} \prod_{j=1; j \neq i}^{k} 1 + |p_{j,h} - p_{j-1,h}| \\
&\leq c \prod_{h=1}^{m} \prod_{j=1; j \neq i}^{k} 1 + |u_{j,\bar{\pi}_h^j} - u_{j-1,\bar{\pi}_h^j - 1}| \\
&\leq c \prod_{h=1}^{m} \prod_{j=1; j \neq i}^{k} 1 + |u_{j,h}|^2
\end{aligned}$$

it is clear that by choosing $\delta > 0$ sufficiently small we can achieve the desired bound.

The second term in (2.12), the t variation, gives rise to a term similar to (2.7) except that for each l; $l = 1, \ldots, m$ the indices $\{i_{1,l}, \ldots, i_{k,l}\}$ run through the set $A = [0, nt]^k - [0, nt']^k$. Using (2.8), we can bound the t variation by a sum over regions $\Delta = \Delta(\pi^1, \ldots, \pi k)$ of integrals of the form (2.11), except that the integrand is replaced by

$$\begin{aligned}
&\frac{1}{n^{km}} \sum_{A^m \cap \Delta} \prod_{j=1}^{k} \prod_{h=1}^{m} |\varphi(u_{j,h}/b_n)|^{\pi_h^j - \pi_{h-1}^j} \\
&\leq (\frac{1}{n^{km}} |A|^m)^{1/q} \{ \prod_{j=1}^{k} \prod_{h=1}^{m} G_{q'}(n, u_{j,h}) \}^{1/q'} \\
&\leq c|t - t'|^{m/q} \prod_{j=1}^{k} \prod_{h=1}^{m} \frac{1}{1 + ((u_{j,h}))^{(\beta - \epsilon)/q'}}
\end{aligned}$$

for any q, q' satisfying $1/q + 1/q' = 1$. It is now clear that by taking q' suficiently close to 1 we can obtain the desired bound on the t variation in (2.12). This completes the proof of our theorem. □

References

[1] R. Bass and D. Khoshnevisan, *Local times on curves and uniform invariance principles*, Preprint.

[2] ———, *Strong approximations to Brownian local time*, Preprint.

[3] A. Borodin, *The asymptotic behavior of local times of recurrent random walks with infinite variance*, Theory Probab. Appl. **29** (1984), 318–333.

[4] E. Perkins, *Weak invariance principles for local times*, Z. Wahrscheinlichkeitstheorie und Verw. Gebiete **60** (1982), 437–451.

[5] J. Rosen, *Random walks and intersection local time*, Ann. Probab. **18** (1990), 959–977.

[6] A.V. Skorohod, *Limit theorems for stochastic processes*, Theory Probab. Appl. **2** (1957), 138–171.

Jay Rosen
Department of Mathematics
The College of Staten Island
City University of New York
New York NY 10301

ON A TWO-ARMED BANDIT PROBLEM WITH BOTH CONTINUOUS AND IMPULSE ACTIONS AND DISCOUNTED REWARDS

A.A.YUSHKEVICH

1. Introduction

In this communication we outline, without full proofs, a computation of the value function and optimal policies in a discounted symmetric Poisson-type two-armed bandit problem (TAB) with both continuous and impulse actions. Our purpose is to present one more physically meaningful example in which an explicit solution of the related quasivariational inequalities (QVI) can be found, and especially, in which optimal policies involve series of impulse decisions instantly following one after another. A simpler, undiscounted version of the same problem is considered by D.S.Donchev [2,3].

Continuous-time multi-armed bandits with Poissonian flows of successes and continuous actions only were introduced originally by E.L.Presman and I.M.Sonin [8]. We refer to the monograph [8] for a detailed discussion of basic concepts utilized here, including a formal reduction of Poisson-type Bayesian bandits, which are controlled processes with incomplete information, to processes of corresponding a posteriori probabilities, which turn to be completely observed controlled piecewise deterministic Markov processes (PDP). Explicit solutions of TAB problems of that type in the case of undiscounted rewards were obtained in [8], and in the case of discounted rewards were found later by E.L.Presman [7] and the author [9].

The QVI approach to impulsively or both impulsively and continuously controlled stochastic processes is due to A.Bensoussan and J.-L.Lions [1]. A rigorous treatment of continuous-time problems with possible immediate repetitions of impulse decisions may be performed in terms of processes with time splitting. QVI-type optimality conditions in those terms for controlled PDP are given in A.A.Yushkevich [10], so that both [2,3] and the present communication are illustrations to [10].

In the Poisson-type TAB problems, as in pioneering studies on the discrete-time TAB (see, for instance, D.Feldman [4]), the arms are dependent in the sense that they are linked by common hypotheses.It should be pointed out that there is an important different direction in the study of continuous-time bandits, with much more general stochastic processes to select from but with independent arms, represented by works of I.Karatzas [5], A.Mandelbaum [6] and others.

In Section 2 we describe the physics of the TAB problem. In Section 3 we reformulate the problem in terms of a posteriori probabilities as an optimization problem for a controlled one-dimensional PDP with time splitting, and explain the way the solution is found. Section 4 contains optimality conditions for such processes. In more technical Sections 5-7 we evaluate explicitly the value function of the model and get stationary optimal policies.

2. The TAB problem

We describe the initial two-armed bandit (TAB) problem on a informal physical level.

An unobserved Poissonian flow of particles of rate 1 comes from the outside and should be distributed by a player (or controller) between the inputs of two arms of a device called TAB. More precisely, at each time $t \geq 0$ the player selects probabilities $a=a_t$ and $1-a$ with which a particle, if arriving at that time, is directed, without being observed by the player, to the first or the second arm respectively. The numbers $a \in [0,1]$ are *continuous actions* of the player (the above name does not mean that a_t is a continuous function of t). Once put into an arm, a particle either vanishes, or instantly passes to the output of this arm where it is observed by the player. Every such observation yields him a unit of reward. The player knows that probabilities of passing are equal to p for one arm and to q for the other, where $0<q<p<1$, but he does not know which of the two arms is better, i.e. which has the greater probability p of passing. To make the model more symmetric, we assume that

$$p+q=1. \tag{1}$$

Besides continuous actions $a\in[0,1]$, the player can use an *impulse action*, a=i. The action i consist in putting a dummy particle into the input of one of the arms (to be definite, say, of the first arm - see Remark (2) below). The sense of this action is to observe whether the particle passes through or not, and in this way to get more information about which of the arms is better. Every utilization of i proceeds instantly and regardless of its outcome has a cost $c\geq 0$. The player can use the action i at any moment, but only a finite number of times on every finite time interval. Moreover, observing the outcome of an impulse action, he can instantly repeat it, and proceed in the same manner any finite number of times before he returns to continuous actions.

Denote by H_1 the hypothesis that probability q of passing is assigned to the first arm and p is assigned to the second, and denote by H_2 the inverse hypothesis. In addition to parameters of the model the player knows *a priori probabilities* x and 1-x of H_1 and H_2, $0<x<1$. *The aim* of the player is to maximize the expected discounted total reward over an infinite horizon.

More formally, denote by N(t) the total number of outside particles observed by the player up to time t at the outputs of both arms, and by I(t) the number of dummy particles utilized by him up to time t. Then the expected reward is equal to

$$w(x,\pi)=E_x^{\pi}\int_0^{\infty}e^{-\alpha t}d[N(t)-cI(t)], \quad 0<x<1,$$

where x is the a priori probability of H_1, π stands for a policy of the player, E_x^{π} denotes the corresponding expectation, and $\alpha>0$ is the discount parameter. Here a policy determines current actions of the player on the basis of past and present observations. *The primary problem is to maximize* $w(x,\pi)$ *over* π.

To simplify solution formulas, we subtract from $w(x,\pi)$ an appropriate constant; evidently, this does not change optimal policies. On the other hand, this gives the criterion another interpretation.

Under any policy π the rate of the process $N(t)$ does not exceed p, therefore certainly $E_x^\pi dN(t) \le pdt$; and this rate is equal to p if the outside particles are directed to the better of the arms. In the latter case, if it were so during the whole time of observations, and if no impulse actions were used, the expected reward would be equal to the integral of $e^{-\alpha t}pdt$, i.e. to $p\alpha^{-1}$. Hence the certainly negative functional $v(x,\pi)=w(x,\pi)-p\alpha^{-1}$, or in an explicit form

$$v(x,\pi)=E_x^\pi\int_0^\infty e^{-\alpha t}[dN(t)-pdt-cdI(t)], \tag{2}$$

has, up to its sign, the sense of losses due to the use of the wrong arm and utilization of the impulse action. Thus a maximization of $w(x,\pi)$ or $v(x,\pi)$ means a minimization of the above losses. With the functional (2) we define *the value function*

$$v(x)=\sup_\pi v(x,\pi), \quad 0<x<1$$

and *optimal policies*: π is optimal if $v(x,\pi)= v(x)$ for all $x\in(0,1)$. In these terms we have

Problem 1 (the TAB problem). Evaluate the value function $v(x)$, $x\in(0,1)$ and find an optimal policy.

Remarks. (1) In the limiting case when $\alpha=0$, i.e. without discounting, the maximization problem with functional w has no sense because $w(x,\pi)=+\infty$ for any reasonable policy. With the functional $v(x,\pi)$ the problem has sense also in the case $\alpha=0$, and is studied in this case by Donchev [2,3].

(2) As noted by Donchev [2], under assumption (1) there is no difference which arm to use for the impulse action. Namely, a success of this action at one arm is equivalent to its failure at the other: both events have the same probability and the same effect on a posteriori probabilities of H_1 and H_2.

3. The optimization problem in terms of a controlled PDP with time splitting

As usual, the initial TAB problem with partial observations is reduced to a control problem with complete information, stated in terms of a posteriori probabilities.

Because $P(H_1)+P(H_2)=1$, it is enough to consider only the probability x_t of H_1 at time t. By applying on an intuitive level the Bayes formula to an infinitesimal time interval (for a formal approach see Presman and Sonin [8]), one easily finds that between observations of particles or utilizations of the impulse action this probability is governed by a controlled deterministic differential equation

(3) $$dx_t=(p-q)(2a_t-1)x_t(1-x_t)dt.$$

Hence $\{x_t\}$ is a controlled piecewise deterministic process (PDP) on the state space $X=(0,1)$. Observations of outside particles passing through the arms under a continuous action $a\in[0,1]$ imply accidental (i.e.Poisson-type) jumps from a current state x either to $x^-=qx[qx+p(1-x)]^{-1}$ or to $x^+=px[q(1-x)+px]^{-1}$ with respective rates $a[qx+p(1-x)]$ and $(1-a)[q(1-x)+px]$. The action i produces an impulse (i.e. an immediate) jump from a current state x to x^- or x^+ with respective probabilities $qx+p(1-x)$ and $q(1-x)+px$.

Mention that in order to deal properly with instant repetitions of impulse actions, we formally understand $\{x_t\}$ as a *controlled process with time splitting* (see [10] for details). In our specific model it means the following. A single current state x_t at time t is defined only on open time intervals between jumps of the process. At a jump epoch t, because of iterations of impulses, the path may have m consecutive jumps (m=1,2,...). To describe the process and the control at this time t, we split it into m+1 consecutive *moments* (t,0),(t,1),...,(t,m). At a moment (t,k) the process is in a current state x(t,k) with $x_{t-}=x(t,0)$, $x(t,m)=x_{t+}$. The current actions are defined at the same times or moments as the current states. Namely, a selection of a_t (or a(t,k)) is based upon an observation of x_t (respectively x(t,k)) and a preceding history. In the case of a splitting into m+1 moments, the action a(t,0) may be either continuous or impulsive (the first jump at t may be of either type), the actions a(t,k) with k=1,2,...m-1 are necessary impulsive, the action a(t,m) is necessary continuous.

According to the sense of the model, for any state $z \in X$ and any continuous action $a \in [0,1]$

$$E_x^\pi[dN(t)-pdt|a_t=a,x_t=z] = $$
$$=[zaq+(1-z)ap+z(1-a)p+(1-z)(1-a)q-p]dt=$$
$$=(q-p)[az+(1-a)(1-z)]dt.$$

Therefore in new terms the functional (2) becomes

$$v(x,\pi)=E_x^\pi\int_0^\infty e^{-at}\{(q-p)[a_t x_t+(1-a_t)(1-x_t)]dt-cdI_t\}$$

(because the number of splitted times t is at most countable, in integrating with respect to dt we may neglect x(t,k) and a(t,k); rewards related to actions a(t,k) are included into the term $-cdI_t$).

As in preceding works on TAB (see Feldman [4] or Presman and Sonin [8]), the picture simplifies in a logarithmic scale

$$(4) \qquad y=\ln x-\ln(1-x), \quad V(y,\pi)=v(x,\pi), \quad x\in(0,1)$$

which arises naturally if one integrates (3), and in which the process $\{x_t\}$ transforms into $\{y_t\}$. Denote

$$(5) \qquad \begin{aligned} &r=p-q, \quad b=pq^{-1}>1, \quad h=\tfrac{1}{2}\ln b>0, \\ &Q(y)=1-P(y)=(p+qe^y)(1+e^y)^{-1}. \end{aligned}$$

Then in the new coordinates: (i) instead of (3) one has

$$(6) \qquad dy_t=r(2a_t-1)dt$$

between jumps of the controlled process $\{y_t\}$; (ii) under an action $a\in[0,1]$, an accidental jump from a current state y occurs, directed to y-2h or y+2h with respective rates aQ(y) and (1-a)P(y); (iii) under the action i at a current state y, an immediate impulse jump occurs, directed to y-2h or to y+2h with respective probabilities Q(y) and P(y); (iv) the criterion becomes

$$(7)\qquad V(y,\pi)=E_y^{\pi}\int_0^{\infty}e^{-\alpha t}[r(y_t,a_t)dt-cdI_t],\qquad y\in R=(-\infty,\infty)$$

with

$$(8)\qquad r(y,a)=-r[(1-a)+ae^y](1+e^y)^{-1},\ a\in[0,1],\ y\in R.$$

The corresponding *value function* is $V(y)=\sup_\pi V(y,\pi)$, and a policy π is *optimal* if $V(y,\pi)=V(y)$ for all y. So we get the following

Problem 2 (Optimization problem for a controlled PDP with time splitting). Evaluate the value function $V(y)$, $-\infty<y<\infty$ and find an optimal policy π.

The stated problem precisely fits into the general framework of controlled processes with both continuous and impulse actions and with time splitting treated in [10]. Formally, we have a time-homogeneous model with an infinite horizon in the sense of [10], say *model Z*, specified by the following elements:

1) A state space E. Here $E=R=(-\infty,\infty)$.

2) An action space A consisting of sets A^1 of impulse actions and A^2 of continuous actions. Here $A^1=\{i\}$ and $A^2=[0,1]$.

3) Sets $A(y)$ of actions admitted at states $y\in E$. Here $A(y)=A$ for all y.

4) A drift function $b(y,a)$, $a\in A^2$, $y\in E$. Here (cf.(i)) $b(y,a)=r(2a-1)=(p-q)(2a-1)$, $a\in[0,1]$, $y\in R$.

5) A jump rate $\mu(y,a)$, $a\in A^2$, $y\in E$. Here (cf.(ii)) $\mu=aQ(y)+(1-a)P(y)$, $a\in[0,1]$, $y\in R$.

6) A jump distribution, i. e. a stochastic kernel $q(y,a,\cdot)$ from ExA into E. Here(cf.(ii) and (iii)) this kernel is defined by conditions $q(\cdot,\cdot,E)=1$, $\mu(y,a)q(y,a,y-2h)=aQ(y)$ and $\mu(y,a)q(y,a,y+2h)=(1-a)P(y)$ if $a\in[0,1]$, $q(y,i,y-2h)=Q(y)$ and $q(y,i,y+2h)=P(y)$.

7) A reward rate $r(y,a)$, $a\in A^2$, $y\in E$. Here (cf.(iv)) it is given in (8).

8) An instantaneous reward function $R(y,a)$, $a\in A^1$, $y\in E$. Here (cf.(iv)) $R(y,i)=-c$, $y\in R$.

9) A discount parameter $\alpha>0$.

Formally in terms of [10], Problem 2 means: *find the value function and an optimal policy in the model Z.*

Any policy π in Problem 2 can be treated as a policy in Problem 1 (based on a continuous computation of current a posteriori probabilities); moreover, if π is optimal in the second problem, then π is optimal in the first one (see Presman and Sonin [8] for details). Therefore and in view of (4), value functions of two problems are related by a formula

$$v(x)=V(\ln x-\ln(1-x)), \quad x\in(0,1),$$

and any solution of Problem 2 is also a solution of Problem 1. Henceforth we deal only with Problem 2.

Solution of Problem 2 is based on sufficient optimality conditions for controlled PDP with time splitting which have a form of quasivariational inequalities (QVI) [10]. These conditions are presented in Section 4 in a simplified one-dimensional version.

In Section 5, utilizing the equality part of QVI, we compute the reward functional $V(y,f_Y)=V(y,Y)$ for a family of stationary policies f_Y depending on a threshold parameter $Y\geq 0$. Selection of those policies is inspired by the sense of impulses and the symmetry of the model. Namely, f_Y assigns the impulse action i to states $y\in[-Y,Y]$ where chances of both hypotheses H_1 and H_2 to be true are relatively high, and assigns the action a=1 (or a=0) to states $y<-Y$ (respectively, $y>Y$) where chances of H_1 (resp., H_2) to be true are essentially higher than those of H_2 (resp., H_1). An exact computation of $V(y,Y)$ becomes possible due to conditioning with respect to H_1 and H_2, after which the process $\{y_t\}$ reduces to three Markov processes, each with constant local characteristics, on the three corresponding parts $(-\infty,Y)$, $[-Y,Y]$ and (Y,∞) of the state space E (on the inner interval it is indeed a Markov chain $y(t,k)$ with $k=0,1,2,\ldots$as a discrete time parameter).

It remains to find the best threshold Y and to check for this Y the inequality part of QVI. A difficulty here is that the dependence of V upon Y is given by different formulas on infinitely many different intervals, and these formulas adjoin in a non-smooth way. Nevertheless it turns out that the global maximum exists, satisfies the first derivative test, and does not depend on y. The corresponding equation relates Y to the cost c of impulses. Existence and uniqueness of its solution follows from a study of the inverse function $c=c(Y)$. All this requires considerable computations presented up to reproducible details in Section 6. But we omit purely technical and lengthy calculations involved in the verification of inequalities (see Section 7 for related and concluding remarks).

4. Stationary policies and the QVI problem

We need a formal definition of a *stationary* policy because it enters into optimality conditions.

Every Borel measurable mapping f of the state space E into the action space A is called a *selector*. A selector f may determine a policy by a rule: apply the action $a=f(y)$ in the state y; in such a case, for simplicity, f itself will be called a policy. But not all selectors do determine policies in the above way. One may show that the following assumptions (A_1)-(A_2) are sufficient for this purpose up to a possible accumulation of impulse decisions. Denote

$$E_i=E_i(f)=\{y\in E: f(y)=i\},\quad E_c=E_c(f)=\{y\in E: f(y)\in[0,1]\}.$$

(A_1) The set E_i is closed.

(A_2) For every $y\in E_c$ the equation

$$y_t=y+r\int_0^t[2f(y_s)-1]ds,\quad t\geq 0 \tag{9}$$

(cf.(6)) has a unique solution up to the first hitting time τ of the set E_i by the path $\{y_t\}$.

The corresponding probability measure on sample paths of the process is constructed by means of the Ionescu Tulcea theorem, by sequentially using (9) on time intervals up to the next jump produced either by continuous or impulse actions. Let $P_y{}^f$ denote this measure for a given f and an initial state y, and let (τ_j,μ_j) be the moments at which the action i is used. Now we can formulate the assumption

(A_3) For every $y\in E$, with $P_y{}^f$-probability 1 either the number of moments (τ_j,μ_j) is finite or $\tau_j\to\infty$ as $j\to\infty$.

Definition 1. A stationary policy is a (policy determined by a) selector f satisfying (A_1)-(A_3).

In the case of a one-dimensional state space E=R we have here, sufficient optimality conditions for controlled PDP with both continuous and impulse actions and with time splitting given in [10] reduce to the following proposition 1.

Define on real-valued functions u(y) on E two dynamic programming operators evidently related to the maximization problem with the reward functional (7):

$$T_i u(y)=-c+Q(y)u(y-2h)+P(y)u(y+2h)-u(y),$$

$$(10)\quad T_a u(y)=r(y,a)+(2a-1)ru'(y)+aQ(y)[u(y-2h)-u(y)]+$$

$$+(1-a)P(y)[u(y+2h)-u(y)]-\alpha u(y),$$

$$y\in R,\quad a\in[0,1]$$

(T_a is defined only at points where the derivative u'(y) exists). Then (cf. [10], Theorem 7.2)

Proposition 1. *If a stationary policy* f *and a bounded locally Lipschitzian function* u *on* E *satisfy conditions*

(C_1) $T_i u(y)\le 0$, $y\in E$,

(C_2) $T_i u(y)=0$, $y\in E_i$,

(C_3) $T_a u(y)\le 0$ for all $a\in[0,1]$, $y\in E$,

(C_4) $T_{f(y)}(y)=0$, $y\in E_c$

then u *is equal to the value function* V *and* f *is optimal.*

Analytically, our purpose is to solve QVI (C_1)-(C_4). Because T_a is linear in a, (C_3) is equivalent to

(C_3') $T_0 u(y)\le 0$ and $T_1 u(y))\le 0$, $y\in E$.

5. Expected rewards under threshold policies

In Sections 5-6 we suppose that c>0. The following *threshold policies* depending on one parameter $Y\ge 0$:

$$(11)\quad f_Y(y)=\begin{cases}1 \text{ if } -\infty<y<-Y,\\ i \text{ if } -Y\le y\le Y,\ (Y\ge 0),\\ 0 \text{ if } Y<y<\infty\end{cases}\qquad g(y)=\begin{cases}1 \text{ if } y<0,\\ \frac{1}{2} \text{ if } y=0,\\ 0 \text{ if } y>0,\end{cases}$$

we consider as candidates for optimal policies (cf. Section 3). These selectors satisfy assumptions (A_1)-(A_3). Namely, g is the known optimal policy in the same model without impulse actions (Presman [7], Yushkevich [9]), and under f_Y the process has the following structure: (1) on the rays $y<-Y$ and $y>Y$ there is a continuous drift towards the origin with a constant velocity r, interrupted by Poisson-type jumps (of a variable length and a variable but bounded rate) in the opposite direction; (2) on the segment $[-Y,Y]$, where the real time stops and splits, there is a discrete-time Markov chain (random walk with a step of length 2h) induced by impulse actions; (3) each time as the continuous-time process reaches $-Y$ or Y, it changes into the chain, and each time as the chain jumps out of $[-Y,Y]$, it changes into the continuous-time process.

To simplify notations, let V_Y stand for $V(\cdot,f_Y)$, and V_g for $V(\cdot,g)$. To compute V_Y, we utilize a following version of the Proposition 1 related to a model with a single action f(y) admitted in the state y.

Proposition 2. *If a stationary policy* f *and a bounded locally Lipschitzian function* u *on* E *satisfy the Conditions* (C_2) *and* (C_4) *of the Proposition* 1, *then* $u(\cdot)=V(\cdot,f)$.

The condition (C_4), in view of (5),(8) and (10), yields for u(y), $y>Y$ an equation

$$[P(y)-p]-ru'(y)+P(y)[u(y+2h)-u(y)]-\alpha u(y)=0.$$

As in Presman and Sonin [8], we observe that

$$u(y)=P_y(H_1)u_1(y)+P_y(H_2)u_2(y)=xu_1(y)+(1-x)u_2(y)=$$
$$=[e^y u_1(y)+u_2(y)](1+e^y)^{-1}$$

where u_1 and u_2 have the same meaning as u, only given H_1, resp. H_2. Under H_1 or H_2 the jump and reward rates are constant, and the differential equations become

$$-ru_1'(y)+p[u_1(y+2h)-u_1(y)]-\alpha u_1(y)=0,$$

$$-r-ru_2'(y)+q[u_2(y+2h)-u_2(y)]-\alpha u_2(y)=0.$$

Their bounded solutions in the domain $y>Y$ are found by standard methods in form of exponential functions. Utilizing also the symmetry between y and -y, we get

$$\alpha(1+e^y)u(y)=\begin{cases}-re^y+Ke^{\beta y} & \text{if } y<-Y,\\ -r+Ke^{-\delta y} & \text{if } y>Y\end{cases} \tag{12}$$

where K is an arbitrary constant and β and δ are the unique positive solutions of equations

$$pb^{-\beta}=p+\alpha-r\beta, \qquad qb^{-\delta}=q+\alpha-r\delta. \tag{13}$$

(b is defined in(5)). It follows that

$$\beta=1+\delta.$$

The condition (C_2) in view of (10) gives an equation

$$u(y)=-c+Q(y)u(y-2h)+P(y)u(y+2h), \quad y\in[-Y,Y]. \tag{14}$$

The same equation is true for the total expected reward in a random walk on the lattice $y\pm 2jh$ $(j=0,1,2,...)$ with transition probabilities $Q(y)$ and $P(y)$ and a reward $-c$ on each step. At points $y-2kh$ and $y+2mh$ outside the interval $[-Y,Y]$ which are closest to it from the left or the right, $u(y)$ is given by (12). Conditioning with respect to H_1 (or H_2) reduces this random walk to two random walks with constant transition probabilities q and p (respectively, p and q). Utilizing the solution of the classical gambler's ruin problem, one may express $u(y)$ for y in $[-Y,Y]$ through the values of u at the above points $y-2kh$ and $y+2mh$. In notations

$$B_j=b^j-1, \qquad B_{km}=kb^{k+m}-(k+m)b^k+m \tag{15}$$

one finally gets

$$\begin{aligned}\alpha rB_{k+m}(1+e^y)u(y)=&-c\alpha(B_{km}+B_{mk}e^y)-r^2(B_k+B_me^y)+\\ &+rK(B_kb^{-m\delta}e^{-\delta y}+B_mb^{-k\delta}e^{\beta y})\end{aligned} \qquad (-Y\le y\le Y) \tag{16}$$

where k and m are functions of y determined by conditions: either

$$(17) \quad 2(k-1)h-Y\le y\le Y+2(m-1)h,\ k+m=n+1,\ k=1,2,\dots,n$$

or

$$(18) \quad Y-2mh<y<2kh-Y,\ k+m=n,\ k=1,2,\dots,n-1$$

and the integer n is a function of Y defined by

$$(19) \quad (n-1)h\le Y<nh.$$

To identify the constant K in (12) and (16), we require from u(y) to be continuous at y=Y (then, because of (14) and the symmetry, u will be continuous also at all other possible break points ±Y±2jh). Utilizing (12) with y=Y and (16) with y=Y, k=n, m=1, we get

$$(20) \quad K=K(Y)=Z^{\delta}[B_1r^2(b^n-Z)-\alpha c(B_{n1}+B_{1n}Z](W_n)^{-1}$$

$$(n-1)h\le Y<nh,\quad n=1,2,\dots$$

where

$$(21) \quad W_n=r(B_{n+1}-B_nb^{-\delta}-B_1b^{-n\delta}Z^{\beta+\delta})>0,$$

$$(22) \quad Z=e^{Y}.$$

With this K the policy f and the function u satisfy all conditions of the Proposition 2.

To compute $u=V_g$, one should use (12) with Y=0 and the boundary condition at y=0:

$$-2r+u(-2h)+u(2h)-2u(0)-4\alpha u(0)=0$$

which follows from (C_4) at y=0, i.e. from $T_{\frac{1}{2}}u(0)=0$ (cf.(11)). This gives the value

$$(23) \quad K=K(g)=r(\beta+\delta)^{-1}.$$

Utilizing the obtained formulas and summarizing we have

Lemma 1. *The expected reward function* V_Y, $Y\geq 0$ *is equal to the function* u *defined by the formulas* (12)-(13) *and* (15)-(22). *This function is bounded, continuous, and, except at points* $\pm Y+2jh\in[-Y,Y]$, $j=0,\pm 1,\pm 2,\ldots$, *has a bounded derivative. The function* V_g *is given by* (12)-(13) *and* (23) *with* Y=0 *and is bounded and everywhere smooth.*

6. The best threshold policy

Now we consider which of the threshold policies maximizes the constant K in (12) and (16).

Lemma 2. *The function* K(Y), $Y\geq 0$ *defined in* (20) *is continuous and*

$$(24)\qquad K(0)=(r^2-2\alpha c)(1-2qb^{-\delta})^{-1},\quad \inf K(Y)=-\infty.$$

Indeed, direct computation shows that

$$(25)\quad K(nh)=(z_n)^{\delta}[B_1r^2z_n-\alpha c(B_{n+1}+(n+1)B_1z_n)](rw_n)^{-1}$$

where

$$(26)\qquad \begin{gathered} w_n=bz_n+1-b^{-\delta}(1+z_n), \\ z_n=e^{nh}=b^{\frac{1}{2}n},\quad n=0,1,2,\ldots, \end{gathered}$$

and that K(nh-0) is equal to (25) for $n\geq 1$. To obtain (24), set in (25) n=0 or $n\to\infty$. QED

From (20)-(22) the derivative of K is equal to

$$(27)\qquad \begin{gathered} K'(Y)=rZ^{\delta}(B_1r^2R_n-\alpha cS_n)(W_n)^{-2}, \\ (n-1)h<Y<nh,\quad n=1,2,\ldots \end{gathered}$$

where W_n is given in (21), $Z=e^Y$ and

$$(28)\qquad \begin{gathered} R_n=(B_{n+1}-B_nb^{-\delta})(\delta b^n-\beta Z)+B_1b^{-n\delta}(\beta b^n-\delta Z)Z^{\beta+\delta}, \\ S_n=(B_{n+1}-B_nb^{-\delta})(\delta B_{n1}+\beta B_{1n}Z)+B_1b^{-n\delta}(\beta B_{n1}+\delta B_{1n}Z)Z^{\beta+\delta}. \end{gathered}$$

In particular, the (right-hand) derivative at Y=0 is

(29) $K'(0)=\alpha[r-(\beta+\delta)c](1-2qb^{-\delta})^{-2}$.

Lemma 3. *At every local extremum point* $Y>0$ *of the function* $K(Y)$ *the derivative* $K'(Y)$ *exists (and therefore is equal to* 0*).*

At possible break points $Y=nh$, $n>0$, left- and right-hand derivatives exist and may be found from (27)-(28). A direct computation shows that

$$B_{n+2}(B_{n+1}-B_nb^{-\delta}-B_1z_n)^2K'(nh-0)=$$
$$=B_n(B_{n+2}-B_{n+1}b^{-\delta}-B_1b^{-\delta}z_n)^2K'(nh+0)$$

where z_n is defined in (26). Thus $K'(nh-0)$ and $K'(nh+0)$ either both vanish, so that $K'(nh)=0$ exists, or both have the same sign. If nh is an extremum point, the latter case is excluded. QED

Lemma 4. *For any* $Y\geq 0$ *the relation* $K'(Y)=0$ *is equivalent to*

(30) $$c=F(e^Y)$$

where $F(Z)$, $1\leq Z<\infty$ *is defined by the formula*

(31) $$F(Z)=B_1r^2R_n(\alpha S_n)^{-1} \text{ if } z_{n-1}\leq Z\leq z_n,\ n\geq 1$$

(see (26),(28)*). The function* $F(Z)$ *is continuous and decreasing on* $[1,\infty)$*, and also*

(32) $$c_0\equiv F(1)=r(\beta+\delta)^{-1}, \quad F(\infty)=0.$$

For Y different from nh the stated equivalence follows from (27); it is extended to $Y=nh$ by an argument similar to that used in Lemma 3. Continuity of F at maybe break points z_n is verified by a direct computation. A straightforward differentiation of (31) implies for any interval $z_{n-1}<Z<z_n$

$$\alpha(S_n)^2F'(Z)=$$
$$=-B_1r^2\beta\delta(B_{n1}+B_{1n}b^n)(B_{n+1}-B_nb^{-\delta}-B_1b^{-n\delta}Z^{\beta+\delta})^2<0,$$

hence F is decreasing. From (31) and (28) one may find

$$c_n \equiv F(z_n)=B_1 r^2 a^{-1} z_n[(\delta b z_n-\beta)+b^{-\delta}(\beta-\delta z_n)]\cdot$$

$$\cdot[\delta B_{n+1,1}+\beta B_{1,n+1}z_n-b^{-\delta}(\delta B_{n1}+\beta B_{1n}z_n)]^{-1}.$$

This gives (32) by setting $n=0$ or $n\to\infty$. QED

Lemma 5. *If* $c<c_0$ *then* $K(g)<K(0)$ *and* $0<K'(0)$, *and the same is true with "<" replaced by "=" or ">".*

This follows from (23),(24),(29) and (32), with utilization of (13) in the case of $K'(0)$. QED

Lemma 6. (i) *If* $0<c<c_0$, *then* $K(Y)$ *is increasing on* $[0,Y^*]$ *and is decreasing on* $[Y^*,\infty)$ *where* Y^* *is a unique solution of the equation* (30); *in this case also* $K(g)<K(0)$ *and* $K'(0)>0$. (ii) *If* $c=c_0$, *then* $K(Y)$ *is decreasing on* $[0,\infty)$, *and in this case also* $K(g)=K(0)$, $K'(o)=0$. (iii) *If* $c>c_0$, *then* $K(Y)$ *is decreasing on* $[0,\infty)$, *and in this case also* $K(g)>K(0)$, $K'(0)<0$.

In the case (i), according to Lemmas 4 and 3, Y^* is the single possible local extremum point of $K(Y)$. Utilizing Lemma 2 we see that either $K(Y)$ has an absolute maximum at the point Y^* or $K(Y)$ is decreasing on $[0,\infty)$. The latter is impossible since $K'(0)>0$ by Lemma 5. In both cases (ii) and (iii) $K(Y)$ has no local extrema on the half-line $Y>0$, and in view of (24) should be decreasing on $[0,\infty)$. The rest follows from Lemma 5. QED

Formulas (12) and (16)-(22) show that the function $u(Y,y)=V_Y(y)$, $Y\geq 0$, $y\in R$ is uniformly continuous in y in any bounded (Y,y)-region. The same formulas together with Lemma 2 show also that $u(Y,y)$ is continuous in Y everywhere except maybe on lines $Y=y\pm 2jh$, $j=0,1,2,\ldots$ where the integers k and m in formulas (16)-(18) may change, or (12) may change with (16). It follows from these two properties that $u(Y,y)$ is continuous in Y on the whole half-plane $Y>0$.

The same formulas imply that, for any fixed y, the function $u(Y,y)$ is increasing or decreasing in Y simultaneously with $K(Y)$ except maybe at points $Y=y\pm 2jh$. Because both K and u are continuous in Y, this isolated

points cannot interrupt the common behavior of u and K. Direct computation shows that u(Y,y) is smooth in y if (30) is fulfilled. Hence we have the following result that specifies the *best threshold policies*:

Theorem 1. (i) *If* $0<c<c_0$ *then the equation* (30) *has a unique solution* $Y^*>0$, *and for this solution the function* $V_Y{*}$ *is smooth and* $V_Y{*}>V_g$, $V_Y{*}>V_Y$ $(Y\geq 0, Y=Y^*)$.

(ii) *If* $c=c_0$ *then the equation* (30) *has a unique solution* $Y=0$, *and* $V_0=V_g>V_Y$ $(Y>0)$.

(iii) *If* $c>c_0$ *then the equation* (30) *has no solutions, and* $V_g>V_Y$ $(Y\geq 0)$.

7. Conclusion

A final highly technical step is to check for the best threshold policy and the corresponding expected reward function u(y) the inequality conditions (C_1) and (C_3') of Proposition 1 which do not directly follow from the equality conditions (C_2) or (C_4). Each of the inequalities divides into cases according to which of the formulas (12), or (16) with (17), or (16) with (18) should be used for u(y), or u(y-2h), or u(y+2h). For undiscounted rewards similar rather lengthy computations based on repeated differentiations etc. are performed in [3]. We omit them here and give the corresponding result in Theorem 2 without a proof.

In the case c=0 it is possible, utilizing repeated impulse actions, to reach arbitrary large absolute values of y without any cost and in this way to maintain the negative reward rate r(y,a) arbitrary close to 0 (see(8)). Thus the value function in this case is identically 0, but evidently no policy provides exactly an expected zero reward.

Theorem 2. (i) *In the case* $c=0$ *there is no optimal policy, and the value function* $V=0$.

(ii) *In the case* $0<c\leq c_0$, *the stationary policy* f_Y *defined in* (11) *is optimal and the value function* $V(y)=u(y)$ *is given by* (5), (12)-(13) *and* (15)-(22), *where* Y *is the unique solution of* (30).

(iii) *In the case* $c\geq c_0$, *the stationary policy* g *defined in* (11) *is optimal and the value function* $V(y)=u(y)$ *is given by the* (12)-(13) *with* $Y=-0$ *and* (23).

Remarks. (1) In the limit as $\alpha \to 0$ (and therefore $\beta \to 1$, $\delta \to 0$), our formulas for the optimal threshold Y, the value function V(y) and the break values c_n, n=0,1,2,... transfer into formulas for the undiscounted case obtained by Donchev [2,3].

(2) It would be nice to prove the sufficiency of threshold policies in some qualitative way, avoiding the above mentioned numerous inequalities.

(3) An open problem is to describe optimal policies in a similar model without the assumption (1).

REFERENCES

[1] A.Bensoussan et J.-L.Lions, *Contrôle impulsionnel et inéquations quasivariationnelles*, Dunod, Paris, 1982.

[2] D.S.Donchev, *The two-armed bandit problem with continuous time in presence of gradual and impulsive controls*, Russian Math.Surveys **45**(1990), #1(271), 200-202.

[3] D.S.Donchev, *On the two-armed bandit problem with both continuous and impulsive actions*, submitted to Steklov Seminar 3 (editors A.N.Shyryaev et al.).

[4] D.Feldman, *Contributions to the "two-armed bandit" problem*, Ann. Math. Stat. **33**(1962),847-856.

[5] I.Karatzas, *Gittins indices in the dynamic allocation problem for diffusion processes*, Ann. Prob. **12**(1984), 173-192.

[6] A.Mandelbaum, *Continuous multi-armed bandits and multiparameter processes*, Ann. Prob. **15**(1987), 1527-1556.

[7] E.L.Presman, *Poissonian version of the two-armed bandit problem with discounting*, Theory Prob. **35**(1990), 307-317.

[8] E.L.Presman and I.M.Sonin, *Sequential control with incomplete data: Bayesian approach*, Academic Press, New York, 1990 (Russian edition 1982).

[9] A.A.Yushkevich, *On the two-armed bandit problem with continuous time parameter and discounted rewards*, Stochastics **23**(1988), 299-310.

[10] A.A.Yushkevich, *Verification theorems for Markov decision processes with controlled deterministic drift and gradual and impulse controls*, Theory Prob. **34**(1989), 474-496.

Existence and uniqueness of the solution of a semilinear stochastic evolution equation on the whole real line

BIJAN Z. ZANGENEH

Abstract

In this paper we will prove the existence and the uniqueness of the solution to the H-valed integral equation

$$X_t = \int_{-\infty}^{t} U(t-s) f_s(X_s) ds + V_t,$$

where H is a real separable Hilbert space. $U(t)$ is a semigroup generated by a strictly negative definite, self-adjoint unbounded operator A such that A^{-1} is compact and f is of monotone type and is bounded by a polynomial, and V_t is a cadlag adapted process.

Introduction

Let H be a real separable Hilbert space with norm $\| \ \|$ and inner product $\langle \ , \ \rangle$. Suppose $(\Omega, \mathcal{F}, \mathcal{F}_t, P)$ is a complete stochastic basis with a right continuous filtration and $\{W_t, t \geq 0\}$ is an H-valued cylindrical Brownian motion with respect to $(\Omega, \mathcal{F}, \mathcal{F}_t, P)$. We follow Yor's definition of cylindrical Brownian motion [7].

Definition 1 *A family of random linear functionals* $\{W_t, t \geq 0\}$ *on* H *is called a* cylindrical Brownian *motion on* H *if it satisfies the following conditions:* (i) $W_0 = 0$ *and* $W_t(x)$ *is* $\mathcal{F}_t$*-adapted for every* $x \in H$.
(ii) *For every* $x \in H$ *such that* $x \neq 0$, $W_t(x)/\|x\|$ *is a one-dimensional Standard Brownian motion.*

Note that cylindrical Brownian motion is not H-valued because its covariance is not nuclear. For the properties of cylindrical Brownian motion and the definition of stochastic integrals with respect to the cylindrical Brownian motion see [7].

Consider the stochastic semilinear equation

$$dX_t = AX_t dt + f_t(X_t)\, dt + dW_t, \tag{1}$$

where A is a closed, self-adjoint, negative definite, unbounded operator such that A^{-1} is nuclear. A *mild solution* of (1) with initial condition $X(0) = X_0$ is the solution of the integral equation

$$X_t = U(t,0)X_0 + \int_0^t U(t-s)f_s(X_s)ds + \int_0^t U(t-s)dW_s, \tag{2}$$

where $U(t)$ is the semigroup generated by A.

Marcus [4] has proved that when f is independent of t and ω and f is uniformly Lipschitz, then the solution of (2) is asymptotically stationary.

To prove this, he studied the following integral equation:

$$X_t = \int_{-\infty}^t U(t-s)f_s(X_s)ds + \int_{-\infty}^t U(t-s)dW_s, \tag{3}$$

where the parameter set of the processes is extended to the whole real line. This motivated us to study the existence of the solution of the more general equation

$$X_t = \int_{-\infty}^t U(t-s)f_s(X_s)ds + V_t, \tag{4}$$

where $U(t)$ is a semigroup generated by a strictly negative definite, self-adjoint unbounded operator A such that A^{-1} is compact and f is of monotone type and is bounded by a polynomial, and V_t is a cadlag adapted process. In this paper we prove the existence and the uniqueness of the solution to (4). In Chapter 5 of [8] we have proved that finite dimensional Galerkin approximations converge strongly to the solution of (4). In Chapter 6 of [8] we have proved under certain conditions that if V_t is a stationary process, then X_t is also a stationary process.

1 Measurability of the Solution

Let $(\Omega, \mathcal{F}, \mathcal{F}_t, P)$ be a complete stochastic basis with a right continuous filtration.

Consider the following integral equation

$$X_t = U(t,0)X_0 + \int_0^t U(t,s)f_s(X_s)ds + V_t, \tag{5}$$

where V_t is a cadlag adapted process and $A = \{A(t), t \in S\}$ is a family of operators satisfying the following hypothesis.

Hypothesis 1 **(a)** *$\exists \lambda \in \mathbf{R}$ such that $\forall s > 0$, $(A(s) - \lambda I)$ is a generator of a contraction semigroup;*
(b) *the operator-valued function $(-A(t) + \mu I)^{-1}$ is strongly continuously differentiable with respect to t for $t \geq 0$ and $\mu > \lambda$;*

(c) *there exists a fundamental solution $U(t,s)$ of the linear equation $\dot{u}(t) = A(t)u(t)$. Moreover, if $u_0 \in H$ and $f \in C(S,H)$, then the strong solution of*

$$\begin{cases} \dot{u}(t) &= A(t)u(t) + f(t) \\ u(0) &= u_0 \end{cases} \tag{6}$$

is given by

$$u(t) = U(t,0)u_0 + \int_0^t U(t,s)f(s)ds. \tag{7}$$

If $u_0 \in D(A(0))$ and $f \in C^1(S,H)$, then (7) is also a strong solution of (6).

Note that an evolution operator which satisfies the above condition is a strong evolution operator (see Curtain [2]).

Remark 1 *Note that Hypothesis 1 holds, for example, if $\{A(t), t \in \mathbf{R}^+\}$ is a family of closed operators in H with domain D independent of t, satisfying the following conditions:*
(i) *considered as a mapping of D (with graph norm) into H, $A(t)$ is C^1 in t on $\mathbf{R}^+$ in the strong operator topology.*
(ii) *If $A(t)^*$ is the adjoint of $A(t)$, then $D(A(t)^*) \subset D$ for all t .*
(iii) *$\exists \lambda \in R$ such that*

$$\langle A(t)x, x\rangle \le \lambda \|x\|^2, \quad \forall x \in D(A(t)), \quad t \in S.$$

Proof: See Browder [1]. Let g be an H-valued function defined on a set $D(G) \subset H$. Recall that g is *monotone* if for each pair $x, y \in D(g)$,

$$\langle g(x) - g(y), x - y\rangle \ \ge 0,$$

and g is *semi-monotone* with *parameter M* if, for each pair $x, y \in D(g)$,

$$\langle g(x) - g(y), x - y\rangle \ \ge -M\|x - y\|^2.$$

We say g is *bounded* if there exists an increasing continuous function ψ on $[0,\infty)$ such that $\|g(x)\| \le \psi(\|x\|), \forall x \in D(g)$. g is *demi-continuous* if, whenever (x_n) is a sequence in $D(g)$ which converges strongly to a point $x \in D(g)$, then $g(x_n)$ converges weakly to $g(x)$.
The following are relevant hypotheses concerning X_0 , f and V.

Hypothesis 2 **(a)** $X_0 \in \mathcal{F}_0$.
(b) *$f = f(t,\omega,x)$ and $V = V(t,\omega)$ are adapted and optional.*
Suppose there exists a set $G \subset \Omega$ such that $P(G) = 1$. If $\omega \in G$, then f and V satisfy **(c)** *for each $t \in S$ and $\omega \in G$, $x \to f(t,\omega,x)$ is demicontinuous and uniformly bounded in t. (That is, there is a function $\varphi = \varphi(x)$ on $\mathbf{R}_+$ which is continuous and increasing in x and such that for all $t \in S$, $x \in H$, and $\omega \in G$, $\|f(t,\omega,x)\| \le \varphi(\|x\|)$.)*

(d) *There exists a non-negative number M such that for each $t \in S$ and $\omega \in G$, $x \to -f(t,\omega,x)$ is semimonotone with parameter M.*
(e) *For each $\omega \in G$, $t \to V(t,\omega)$ is cadlag.*

Proposition 1 *Suppose that X_0, f and V satisfy Hypothesis 2. Suppose A and U satisfy Hypothesis 1. Then (5) has a unique adapted cadlag (continuous, if V_t is continuous) solution. Furthermore*

$$\|X(t)\| \leq \|X_0\| + \|V(t)\| + \int_0^t e^{(\lambda+M)(t-s)}\|f(s,U(s,0)X_0+V(s))\|ds, \quad (8)$$

$$\|X\|_\infty \leq \|X_0\| + \|V\|_\infty + C_T\varphi(\|X_0\| + \|V\|_\infty), \quad (9)$$

where

$$C_T = \begin{cases} \frac{1}{M+\lambda}e^{(M+\lambda)T} & \text{if } M+\lambda \neq 0 \\ 1 & \text{otherwise.} \end{cases}$$

If X_1 and X_2 are solutions corresponding to different initial values X_{01} and X_{02}, then

$$\|X_2(t) - X_1(t)\| \leq e^{(\lambda+M)t}\|X_{01} - X_{02}\|, \quad t \in S. \quad (10)$$

Proof: See [9] or [8] Corollary 4.1.

In [10], the measurablity of the solution of (5) is used to prove the existence of the solution of the stochastic semilinear integral equation

$$X_t = U(t,0)X_0 + \int_0^t U(t,s)f_s(X_s)ds + \int_0^t U(t,s)g_s(X)dW_s + V_t,$$

where
• $g_s(.)$ is a uniformly-Lipschitz predictable functional with values in the space of Hilbert-Schmidt operators on H; • $\{W_t, t \in \mathbf{R}\}$ is an H-valued cylindrical Brownian motion with respect to $(\Omega, \mathcal{F}, \mathcal{F}_t, P)$.

Examples

1. Example.

The existence and uniqueness of the solution of the integral equation (2) have been studied in Marcus [5]. He assumed that f is independent of $\omega \in \Omega$ and $t \in S$ and that there are $M > 0$, and $p \geq 1$ for which

$$\langle f(u) - f(v), u - v \rangle \leq -M\|u-v\|^p$$

and

$$\|f(u)\| \leq C(1 + \|u\|^{p-1}).$$

He proved that this integral equation has a unique solution in $L^p(\Omega, L^p(S,H))$. As a consequence of Proposition 1 we can extend Marcus' result to more general f and we can show the existence of a strong solution of (2) which is continuous instead of merely being in $L^p(\Omega, L^p(S,H))$.

The Ornstein–Uhlenbeck process $V_t = \int_0^t U(t-s)dW(s)$ has been well-studied e.g. in [3] where they show that V_t has a continuous version. We can rewrite (2) as

$$X_t = U(t)X(0) + \int_0^t U(t-s)f_s(X_s)ds + V_t,$$

where V_t is an adapted continuous process. Then by Proposition 1 the equation (2) has a unique continuous adapted solution.

2. Example.

Let D be a bounded domain with a smooth boundary in $\mathbf{R}^d$. Let $-A$ be a uniformly strongly elliptic second order differential operator with smooth coefficients on D. Let B be the operator $B = d(x)D_N + e(x)$, where D_N is the normal derivative on ∂D, and d and e are in $C^\infty(\partial D)$.

Let A (with the boundary condition $Bf \equiv 0$) be self-adjoint.

Consider the initial-boundary-value problem

$$\begin{cases} \frac{\partial u}{\partial t} + Au &= f_t(u) + \dot{W} \quad \text{on} \quad D \times [0,\infty), \\ Bu &= 0 \quad \text{on} \quad \partial D \times [0,\infty), \\ u(0,x) &= 0 \quad \text{on} \quad D, \end{cases} \tag{11}$$

where $\dot{W} = \dot{W}(t,x)$ is a white noise in space–time for the definition and properties of white noise see J.B. Walsh [6], and f_t is a non-linear function that will be defined below. Let $p > \frac{d}{2}$. W can be considered as a *Brownian motion* $\tilde{W}_t$ *on the Sobolov space* H_{-p} see Walsh [6], Chapter 4. page 4.11. There is a complete orthonormal basis $\{e_k\}$ for H_p. The operator A (plus boundary conditions) has eigenvalues $\{\lambda_k\}$ with respect to $\{e_k\}$ i.e. $Ae_k = \lambda_k e_k$, $\forall k$. The eigenvalues satisfy $\Sigma_j(1+\lambda_j^{-p}) < \infty$ if $p > \frac{d}{2}$ see Walsh [6], Chapter 4, page 4.9. Then $[A^{-1}]^p$ is nuclear and $-A$ generates a contraction semigroup $U(t) \equiv e^{-tA}$. This semigroup satisfies Hypothesis 1.

Now consider the initial-boundary-value problem (11) as a semilinear stochastic evolution equation

$$du_t + Au_t dt = f_t(u_t)dt + d\tilde{W}_t \tag{12}$$

with initial condition $u(0) = 0$, where $f : S \times \Omega \times H_{-p} \to H_{-p}$ satisfies Hypotheses (2b) and (2c) relative to the separable Hilbert space $H = H_{-p}$. We can define the mild solution of (12) (which is also a mild solution of (11)), to be the solution of

$$u_t = \int_0^t U(t-s)f_s(u_s)ds + \int_0^t U(t-s)d\tilde{W}_s. \tag{13}$$

Since $\tilde{W}_t$ is a continuous local martingale on the separable Hilbert space H_{-p}, then $\int_0^t U(t-s)d\tilde{W}_s$ has an adapted continuous version see for example [10].

If we define

$$V_t := \int_0^t U(t-s)d\tilde{W}_s,$$

then by Proposition 1, equation (13) has a unique continuous solution with values in H_{-p}.

2 A Semilinear Integral Equation on the Whole Real Line

Let us reduce the integral equation (4) to the following integral equation

$$X_t = \int_{-\infty}^t U(t-s)f(X_s + V_s)ds. \tag{14}$$

We will impose the following conditions on f, V and the generator A of the semigroup U.

Hypothesis 3 (a) *$U(t)$ is a semigroup generated by a strictly negative definite, self-adjoint unbounded operator A such that A^{-1} is compact. Then there is $\lambda > 0$ such that $\|U(t)\| \leq e^{-\lambda t}$.*
(b) *Let $\varphi(t) = K(1 + t^p)$ for some $p > 0$, $K > 0$. $-f$ is a monotone demicontinuous mapping from H to H such that $\|f(x)\| \leq \varphi(\|x\|)$ for all $x \in H$.*
(c) *Let $r = 2p^2$. V_t is cadlag adapted process such that $\sup_{t\in\mathbf{R}} E\{\|V_t\|^r\} < \infty$.*

The following theorem translates Proposition 1 to the case when parameter set of the process is the whole real line.

Theorem 1 *If A, f and V satisfy Hypothesis 2, then the integral equation (14) has a unique continuous solution X such that*

$$\|X_t\| \leq \int_{-\infty}^t e^{-\lambda(t-s)}\varphi(\|V_s\|)ds; \tag{15}$$

$$E\{\|X_t\|\} \leq \frac{1}{\lambda}\sup_{s\in R} E\{\varphi(\|V_s\|)\} := K_1. \tag{16}$$

Proof of the Theorem

- **Existence:**

Consider a sequence of solutions (X_n) of the integral equation

$$X_n(t) = \int_{-n}^t U(t-s)f(X_n(s) + V_s)ds. \tag{17}$$

The solution of (17) exists by Proposition 1. It satisfies

$$\|X_n(t)\| \le \int_{-n}^{t} e^{-\lambda(t-s)}\varphi(\|V_s\|)ds \le \int_{-\infty}^{t} e^{-\lambda(t-s)}\varphi(\|V_s\|)ds \text{ for } t > -n.$$

Since by Hypothesis 3 $\sup_{s\in R} E\{\varphi(\|V_s\|)\} < \infty$, then by Fubini's theorem

$$E\{\int_{-\infty}^{t} e^{-\lambda(t-s)}\varphi(\|V_s\|)ds\} = \int_{-\infty}^{t} e^{-\lambda(t-s)}E\{\varphi(\|V_s\|)\}ds.$$

Then

$$E\{\|X_n(t)\|\} \le K_1 \quad \text{for} \quad t > -n. \tag{18}$$

But

$$X_{n+1}(t) = \int_{-n-1}^{t} U(t-s)f(X_{n+1}(s) + V_s)ds,$$

so by using the semigroup property of $U(t)$ we can rewrite this as

$$X_{n+1}(t) = U(t+n)X_{n+1}(-n) + \int_{-n}^{t} U(t-s)f(X_{n+1}(s) + V_s)ds. \tag{19}$$

But (19) is the same equation as (17) with different initial conditions. Then by Proposition 1 we have

$$\|X_{n+1}(t) - X_n(t)\| \le e^{-\lambda(t+n)}\|X_{n+1}(-n) - 0\|,$$

or

$$\|X_{n+1} - X_n\|_T \le e^{\lambda T}e^{-\lambda n}\|X_{n+1}(-n)\|,$$

where $T \ge 0$ is fixed and $\|X\|_T = \sup_{-T\le t\le T}\|X_t\|$. Then

$$E\{\|X_{n+1} - X_n\|_T\} \le e^{\lambda T}e^{-\lambda n}E\{\|X_{n+1}(-n)\|\}.$$

But by (18) $E\{\|X_{n+1}(-n)\|\} \le K_1$, so

$$\sum_{k=0}^{\infty} E\{\|X_{n+1} - X_n\|_T\} \le K_1 e^{\lambda T}\sum_{k=0}^{\infty} e^{-\lambda k} < \infty.$$

Thus (X_n) is a convergent sequence in $L^1(\Omega, C([-T,T], H)$ for each $T > 0$. Define $X = \lim_{n\to\infty} X_n$. Since $X_n(t)$ satisfies (15) and(16) and since $E\{\|X_n(t) - X(t)\|\} \to 0$ then X also satisfies (15) and (16).

To complete the proof of existence we need to show that X satisfies (14). Now we can rewrite equation (17) as

$$X_n(t) = U(t+T)X_n(-T) + \int_{-T}^{t} U(t-s)f(X_n(s) + V_s)ds. \tag{20}$$

Consider the integral equation

$$Y(t) = U(t+T)X(-T) + \int_{-T}^{t} U(t-s)f(Y(s)+V_s)ds. \qquad (21)$$

By Proposition 1 this equation has a unique solution. Comparing (21) with (20), we have by Proposition 1 that

$$\|X_n(t) - Y(t)\| \leq e^{-\lambda(t+T)}\|X_n(-T) - X(-T)\|.$$

Now

$$\begin{aligned} E[\|Y(t) - X(t)\|] &\leq E[\|X(t) - X_n(t)\|] + E[\|X_n(t) - Y(t)\|] \\ &\leq E[\|X(t) - X_n(t)\|] + e^{-\lambda(t+T)}E[\|X_n(-T) - X(-T)\|], \end{aligned}$$

and since $E[\|X(t) - X_n(t)\|] \to 0$ as $n \to \infty$ then we have $X(t) = Y(t)$ a.s., i.e., $X(t)$ is a solution of (21). We can rewrite (21) as

$$X(t) = U(t+n)X(-n) + \int_{-n}^{t} U(t-s)f(X_s+V_s)ds. \qquad (22)$$

By (15), (16) and Hypothesis 3(b) we have

$$\begin{aligned} \|U(t-s)f(X_s+V_s)\| &\leq e^{-\lambda(t-s)}\varphi(\|X_s+V_s\|) \\ &\leq 2^p K_1 e^{-\lambda(t-s)}\left\{1 + \left(\int_{-\infty}^{s} e^{-\lambda p(s-u)}\|V_u\|^p du\right)^p + \|V_s\|^p\right\} \end{aligned}$$

Then by Hypothesis 3(c) and Fubini's theorem it is easy to see that

$$\int_{-\infty}^{t} \|U(t-s)f(X_s+V_s)\|ds < \infty.$$

By the dominated convergence theorem we have

$$\lim_{n\to\infty} \int_{-n}^{t} U(t-s)f(X_s+V_s)ds = \int_{-\infty}^{t} U(t-s)f(X_s+V_s)ds.$$

Since X satisfies (16) then $E[\|X(-n)\|] \leq K$, the

$$E[\|X(t+n)X(-n)\|] \leq e^{-\lambda(t+n)}K_1,$$

which implies that $U(t+n)X(-n) \to 0$ and so (22) implies

$$X_t = \int_{-\infty}^{t} U(t-s)f(X_s+V_s)ds.$$

- **Uniqueness:**

Let X and Y be two adapted continuous solutions of (14). Then they satisfy the integral equation (22) for $n \in \mathbf{N}$ and by (10) of Proposition 1 one has

$$\|X(t) - Y(t)\| \le e^{-\lambda(t+n)}\|X(-n) - Y(-n)\|.$$

But by inequality (16)

$$E(\|X(-n) - Y(-n)\|) \le 2K_1;$$

therefore

$$E(\|X(t) - Y(t)\|) \le 2K_1 e^{-\lambda(t+n)},$$

and so by a limiting argument $X(t) = Y(t)$ a.s. Q.E.D.

Acknowledgement

The author wishes to express his gratitude to Professor J.B. Walsh for his advice, guidance and encouragement.

References

[1] **Browder, F.E.**, Non-linear equations of evolution. *Ann. of Math.*, 80 (1964) , 485-523

[2] **Curtain, R.F.**, Stochastic evolution equations with general white noise disturbance. *Journal of Mathematical Analysis and Applications*, 60 (1977), 570-595.

[3] **Iscoe, I., Marcus, M.B., McDonald, D., Talagrand, M. & Zinn, J.** , Continuity of l^2-valued Ornstein-Uhlenbeck Process. *The Annals of Probability* 18(1) (1990), 68-84.

[4] **Marcus, R.** , Parabolic Ito equations. *Trans. Amer. Math. Soc.*, 198 (1974), 177-190.

[5] **Marcus, R.** , Parabolic Ito equations with monotone non-linearities. *Functional Analysis*, 29 (1978), 275-286.

[6] **Walsh, J.B.** , An introduction to stochastic partial differential equations. *Lecture Notes in Math.*, 1180 (1986), 266-439.

[7] **Yor, M.** , Existence et unicité de diffusions à valeurs dans un espace de Hilbert. *Ann. Inst. H. Poincaré*, 10 (1974), 55-88.

[8] **Z. Zangeneh, B.**, Semilinear Stochastic Evolution Equations, Ph.D thesis, *University of British Columbia* Vancouver, B.C. CANADA 1990.

[9] **Z. Zangeneh, B.** , Measurability of the solution of a semilinear evolution equation. *Seminar on Stochastic Processes, 1990, Birkhäuser*, Boston.

[10] **Z. Zangeneh, B.** , Semilinear stochastic evolution equations with monotone nonlinearities. To appear in *Stochastics and Stochastics Reports.*

Bijan Z. Zangeneh
Department of Mathematics
University of British Columbia
Vancouver, B.C. V6T1Y4
CANADA

Progress in Probability

Progress in Probability is designed for the publication of workshops, seminars and conference proceedings on all aspects of probability theory and stochastic processes, as well as their connections with and applications to other areas such as mathematical statistics and statistical physics. It acts as a companion series to *Probability and Its Applications,* a context for research level monographs and advanced graduate texts.

We encourage preparation of manuscripts in some form of TeX for delivery in camera-ready copy, which leads to rapid publication, or in electronic form for interfacing with laser printers or typesetters.

Proposals should be sent directly to the editors or to:
Birkhäuser Boston, 675 Massachusetts Avenue, Cambridge, MA 02139, U.S.A.

1 ÇINLAR/CHUNG/GETOOR. Seminar on Stochastic Processes, 1981
2 KESTEN. Percolation Theory for Mathematicians
3 ASMUSSEN/HERING. Branching Processes
4 CHUNG/WILLIAMS. Introduction to Stochastic Integration
5 ÇINLAR/CHUNG/GETOOR. Seminar on Stochastic Processes, 1982
6 BLOOMFIELD/STEIGER. Least Absolute Deviation
7 ÇINLAR/CHUNG/GETOOR. Seminar on Stochastic Processes, 1983
8 BOUGEROL/LACROIX. Products of Random Matrices with Application to Schrödinger Operator
9 ÇINLAR/CHUNG/GETOOR. Seminar on Stochastic Processes, 1984
10 KIFER. Ergodic Theory of Random Transformations
11 EBERLEIN/TAQQU. Dependence in Probability and Statistics
12 ÇINLAR/CHUNG/GETOOR. Seminar on Stochastic Processes, 1985
13 ÇINLAR/CHUNG/GETOOR/GLOVER. Seminar on Stochastic Processes, 1986
14 DEVROYE. A Course in Density Estimation
15 ÇINLAR/CHUNG/GETOOR/GLOVER. Seminar on Stochastic Processes, 1987
16 KIFER. Random Perturbations of Dynamical Systems
17 ÇINLAR/CHUNG/GETOOR/GLOVER. Seminar on Stochastic Processes, 1988
18 ÇINLAR/CHUNG/GETOOR/FITZSIMMONS/ WILLIAMS. Seminar on Stochastic Processes, 1989
19 ALEXANDER/WATKINS. Spatial Stochastic Processes: A Festschrift in Honor of Ted Harris on His 70th Birthday
20 HAAGERUP/HOFFMANN-JØRGENSEN/ NIELSEN. Probability in Banach Spaces 6: Proceedings of the Sixth International Conference, Sandbjerg, Denmark 1986

21 EBERLEIN/KUELBS/MARCUS. Probability in Banach Spaces 7: Proceedings of the Seventh International Conference

22 PINSKY. Diffusion Processes and Related Problems in Analysis, Volume I: Diffusions in Analysis and Geometry

23 HAHN, MASON, WEINER. Sums, Trimmed Sums and Extremes

24 ÇINLAR. Seminar on Stochastic Processes, 1990

25 CAMBANIS/SAMORODNITSKY/ TAQQU. Stable Processes and Related Topics

26 CRUZEIRO/ZAMBRINI. Stochastic Analysis and Applications: Proceedings of the 1989 Lisbon Conference

27 PINSKY/WIHSTUTZ. Diffusion Processes and Related Problems in Analysis, Vol. II. Stochastic Flows

28 DURRETT/KESTEN. Random Walks, Brownian Motion and Interacting Particle Systems

29 ÇINLAR/CHUNG/SHARPE. Seminar on Stochastic Processes, 1991

30 DUDLEY/HAHN/KUELBS. Probability in Banach Spaces, 8, 1992

31 KÖREZLIOĞLU/ÜSTÜNEL. Stochastic Analysis and Related Topics, 1992

32 NUALART/SOLE. Barcelona Seminar on Stochastic Analysis: St. Feliu de Guixols, 1991

33 ÇINLAR/CHUNG/SHARPE. Seminar on Stochastic Processes, 1992

MIX
Papier aus verantwortungsvollen Quellen
Paper from responsible sources
FSC® C105338

If you have any concerns about our products,
you can contact us on
ProductSafety@springernature.com

In case Publisher is established outside the EU,
the EU authorized representative is:
Springer Nature Customer Service Center GmbH
Europaplatz 3, 69115 Heidelberg, Germany

Printed by Libri Plureos GmbH
in Hamburg, Germany